THE LANDSCAPE BELOW

THE LANDSCAPE BELOW

Soil, soul and agriculture

Bruce C Ball

wild goose
publications

 www.**ionabooks**.com

First published 2015 by
Wild Goose Publications, Fourth Floor, Savoy House,
140 Sauchiehall Street, Glasgow G2 3DH, UK,
the publishing division of the Iona Community.
Scottish Charity No. SC003794. Limited Company Reg. No. SC096243.

ISBN 978-1-84952-321-9

Overseas distribution
Australia: Willow Connection Pty Ltd, Unit 4A, 3–9 Kenneth Road,
Manly Vale, NSW 2093
New Zealand: Pleroma, Higginson Street, Otane 4170,
Central Hawkes Bay
Canada: Bayard Distribution, 10 Lower Spadina Ave., Suite 400,
Toronto, Ontario M5V 2Z

Printed by Bell & Bain, Thornliebank, Glasgow

CONTENTS

Touch the earth lightly, use the earth gently,
nourish the life of the world in our care:
gift of great wonder, ours to surrender,
trust for the children tomorrow will bear

ACKNOWLEDGEMENTS

Thanks to my daughter for persuading me to write this book. Elaine Walker guided me on the path of creative writing and writing poetry. Mary Norton gave invaluable advice in straightening out my ideas. I am thankful to my colleagues Bob Rees, Tom Batey, Per Schjønning, Mike O'Sullivan, David McKenzie, Willie Towers and Rattan Lal for reading material and/or providing encouragement. Many others also provided comments, advice and support, particularly colleagues from Scotland's Rural College. Alastair McIntosh wrote the Foreword and also provided advice and ideas. Research and development work on soil was supported by several projects sponsored by the Scottish Office, the Scottish Government and the UK Department for Environment, Food and Rural Affairs.

Most of all, I am grateful to the farmers of Clatt for sharing their thoughts, ideas, wisdom and stories, particularly Brian and Duncan Muirden and James Petrie. I am indebted to Brian Muirden for providing access to soils and to community events. I hope that I have been accurate and faithful in my recollections and reporting. I am greatly indebted to my wife Louise for putting up with my endless editing and demands to read and re-read drafts.

For permission to reproduce copyright material the publisher gratefully acknowledges the following: Hope Publishing for the epigraph verse, Sheena Blackhall for 'The Spik o' the Lan!' and Richard Bly for 'The Earthworm' by Harry Edmund Martinson originally published in *Friends, you drank some darkness*.

For permission to reproduce photographs, the publisher gratefully acknowledges: Dr Hubert Boizard, INRA Estrées-Mons, France for Figure 6; Dr Marinus Brouwers, Clapiers, France for Figure 9; Dr Gabriela Brändle, Agroscope, Zurich, Switzerland for Figures 8 and 12, Dr Everton Blainski for Figure 14 and Dr Neyde Fabíola, Ponta Grossa, Brazil for Figure 16.

Thanks to the staff of Wild Goose Publications for their support, particularly to Sandra Kramer for helping to clarify the meaning of the text.

FOREWORD

by Alastair McIntosh

This book is an autobiography of the soil. It is written autobiographically by a son of the soil, an Aberdeenshire ploughman who became a soil scientist and who now finds himself reflecting on the parish soil out of which human beings grow.

Bruce Ball was raised amongst the farmers of Clatt near Huntly in central Aberdeenshire. He grew up in an era when most farms were modernising, but there remained enough of the old ethos for him to have known farmers who worried that nitrogen fertilisers would 'suck the ground' of its goodness, and for whom job satisfaction was a field immaculately ploughed.

His early experience of soil science was through the senses. The crumble and smell of a soil – some would even add its taste – that determines real quality as distinct from the narrow utilitarianism of intensive farming that treats soil merely as a sand and clay medium for hydroponics.

The longer that Bruce meditated on the nature of ground, the more he came to see that we, too, are like the earthworm. We, too, are organisms of the soil, creatures of the parish that is our place on Earth. As I read through this lyrical book, my mind kept going back to Genesis 27:27 where the elderly Isaac says, 'Ah, the smell of my son is like the smell of a field that the Lord has blessed.'

It is true that Isaac was being tricked at the time. In the same way, we have tricked ourselves (or been tricked) into thinking that chemical technology, based on plentiful cheap oil, has done away with the need to cherish soil quality. But Isaac's central point remains deeply valid. As every traditional farmer and many a gardener can testify, the smell of a wholesome human being and the smell of the goodness of the Earth have much in common. If we fail to care for the soil, then what will become of us?

Such is the deeper question that Bruce's writing poses as this book advances towards its spiritual conclusion. To abuse the soil in the way that so much of our modern agriculture is doing is to foreclose on the options and food-supply resilience of future generations. As has been prophetically said: 'Those who destroy their soils ultimately destroy themselves.'

And the resolution? Bruce emphasises the various farming options for reaching food security, but as he sees it, we need nothing less than a transfiguration of our relationship to the soil that makes up the parish. Transfiguration, he says:

> '… *recognises the importance of people, their motivation and their spirituality. It demands a permanent spiritual step-change and a renewal of mind that I believe will allow more of us to become people of the soil, whose inner life or inner soil is grounded in the earth, who know it as their ally, and whose actions reveal their connections with the earth, with others and with the environment. Such people have inner awareness and are often called the salt of the earth.*'

It was not just the metaphor of salt that came to my mind as I read this. I also thought of how Jesus healed the blind man, mixing the spittle from his own body with the soil that he scraped from the ground. That man was, perhaps, physically blind, but the power of the gospel story is how it awakens us to our spiritual blindness, and in such a visceral manner.

Bruce's vision of a transfigured world is one where we might 'live *more simply, but with inner richness, like soils dark with organic matter.*' A world in which, 'like high-quality soil, we grow deeper and *closer together, promoting "us" rather than "me".*'

From such depth of grounding the churches themselves might 'start afresh', as Bruce cites one of the Clatt farmers as suggesting, and so while conceding his uncertainty as to what the future will look like, he feels sure *'that when we make the transfigured world, we will be standing on soil that is resilient, dark and porous, rich and deep with organic matter, wisdom and love.'*

As such, our communities, our membership one of another on the Vine of Life, will be nothing less than *'beauty that is soil deep'*, and if Bruce Ball's prescription is followed, that will be a depth that grows as the Earth itself ages, and the human spiritual journey melts progressively into the fullness of divine experience.

Alastair McIntosh was appointed Scotland's first Visiting Professor of Human Ecology at the University of Strathclyde and is a Fellow of the Centre for Human Ecology, now based in Govan. His books include *Soil and Soul* (Aurum), *Rekindling Community* (Schumacher Briefings), *Hell and High Water: Climate Change, Hope and the Human Community* (Birlinn) and *Parables of Northern Seed* (Wild Goose). His writing has been described by the Bishop of Liverpool as 'life-changing', by the Archbishop of Canterbury as 'inspirational', and by Thom Yorke, the lead singer of *Radiohead*, as 'truly mental'. Like Bruce Ball, he is a graduate of the University of Aberdeen.

INTRODUCTION

Man – despite his artistic pretensions, his sophistication and
his many accomplishments – owes his existence to a 15cm
layer of topsoil and the fact that it rains.
Anon.

I have appreciated the beauty of the natural world all my life.
Growing up in a rural environment, the way I saw the world was
shaped by the hills and rivers and the cycles of flowers, trees and
animals. However, it wasn't until middle age that I recognised the
hidden beauty, in people, voices, eyes, gestures, poetry. That we
often obscure beauty beneath ugliness and the materialism of the
modern world is something I have come to see as a coping strategy
for our loss of connection to the unseen value of all around us. The
challenges society faces today suggest an addiction that has led to
shortages of resources like oil, food and water and to reliance on
things we have come to see as essential, while we ignore the elements of life that have real value. The inequalities evident between
rich and poor can load relationships with guilt and resentment,
causing us to overlook the fact that we are all – in essence – the
same.

I grew up with shortages because my folks didn't have much
money and we lived in a house without a reliable water supply. I
came with economy built in. Our house was isolated but we had
some land, a few hens, cows next door and plenty of wildlife, so
what I lacked in toys and company I made up for by spending time
outside learning the language of the hens, the animals, the birds,
the trees, the vegetables, the stars and the soil.

Recently, I have begun to get impatient with our society which
uses too much of everything, including other people, and which
seems to be so angry and unhappy. Our destructive way of living,
which burns up our last reserves of oil instead of conserving them

for the next generation and which throws away so many things, is plain stupid. Sucking so much water out of the ground and fighting each other is plain stupid, as is our continued degradation of the environment.

I became a soil scientist mainly to learn how crops grew and to discover how to get them to grow better. We rely on the soil to feed us, despite our high-tech society, and we use crops and cultivation techniques that have changed little over the years. Only about 11% of the earth's surface is covered by soil. Yet I soon realised that, even when we think we're caring for it, we often treat that soil like dirt. We chuck fertiliser or water at it, churn it up and squash it back together with bigger and bigger machines. Our soils are losing their quality, becoming exhausted, so that over one third of the areas of productive soils are degraded. The soil is literally and figuratively the foundation of the environment, albeit hidden most of the time. Moreover, our soils are actually disappearing. As you read this, soils are being lost worldwide at an average rate of 300 tons per minute. If this continues, our remaining topsoil might not survive much beyond fifty years.

Shortages of soil, fertiliser and water will make it more and more difficult to feed our increasing population, leading to the very real prospect of food scarcity. Even if we're not greatly concerned about global warming and peak oil, we can't do without food. Nor can we deny that it is happening as some do with global warming. The evidence is clear in the lost soil, in problems with excessive nitrogen fertiliser use, in disappearing freshwater resources, destroyed forests, extinct animal species and desertification. There is no room for 'abused land' sceptics. We have already seen food riots and protests as a result of high prices. We need to learn the lesson from previous civilisations like that of the Maya. We tend to associate them with magnificent temples rising from the jungle. But they lost most of their soil due to erosion under their intensive agriculture. Those who destroy their soils ultimately destroy themselves.

I've spent time in my career developing simple methods to observe soil and to learn from it. When I handle the soil, it says good things to me. It speaks of its willingness to suck up water, to take in fertiliser, to become warm and to feed me. But like everything it has its breaking point, although it also responds to kindness. There is a parallel between caring for the soil and caring for others. I believe that achievement of this caring involves a search for beauty, not as an abstract concept but as a recognition of value.

My search began back in the community where I grew up. The people there loved the soil and conserved the environment instinctively. Most had successfully assimilated modern ways, yet had kept a traditional approach in their relationship with the land and with each other.

As a child, I recognised beauty in the yellow and brown colours of the local soil. As an adult, I discovered the strange purple soils derived from lava-like pumice stone in the nearby village of Gartly. The differences in colour alone remind us of the hidden complexity in the earth we walk on. Soil was described by Hildegard of Bingen four hundred years ago as 'afire with the light of God'. This light comes from below as a crop germinates and emerges – she calls it 'the greening power of the soil'. But the action of this light for growth depends upon the health of the soil and, because of its hidden nature, when its quality diminishes, we may not notice until it is too late.

In a similar way, our mental and spiritual health is being eroded from the foundations of our communities and our wider society. Globally we are facing an economic recession. In the northern hemisphere we are also facing a social recession. Despite the accumulation of material wealth, which we believed would bring us security, there is growing violence, pressure on the environment and widespread social inequality which makes us feel increasingly unhappy and isolated.

My career as a soil scientist began as a search for ways to improve yields for economic growth. My work with church groups began as a means of re-creating community. Both have brought unexpected results. Through recognising the beauty of the soil and the bonds between people, I have a greater appreciation of both and have found that the original aims of economy and community can be achieved in more holistic ways, by reconnection with the land and tuning in to the urge of the spiritual. We need to get back to our roots in the soil and delve within us to access the stored knowledge of past generations. There are close links between how the soil works, and how we manage it, and how we cope with others and manage our lives. This was well illustrated in the community of Clatt where I grew up. The people of the soil – those who not only work the land but also feel for it – were those who fostered community developments. There was a tangible link between caring for the soil and caring for people.

Good soil management is not easy and our quick-fix approaches often lead to long-term damage. We could possibly feed ourselves without environmental damage if we all behaved rationally. But we have never behaved rationally before, so it's unlikely that we will now. We claim to be ecological in fighting pollution and resource depletion. Yet this is often with the objective of preserving the health and affluence of people in developed countries – 'shallow ecology' in other words. In contrast, deep ecology[1] sees the world as a network of phenomena that are all fundamentally linked and interdependent. This approach values all people, animals, plants and minerals equally. To move towards this deep ecological approach I think we need some help from the spiritual. We need to become aware of the sacredness of our environment, to listen and learn and become conscious of our connection to the spiritual within all things, so that we treat everything with reverence. In this book I look to the inner space both of soil and of humanity. I show the urgent importance of soil life for the future of humankind. The management of soils and farming can integrate with the spiritual and the social to point us towards the creation of sustainable communities based on beauty, justice and transformation. In this way we will be able to satisfy our needs and desires without robbing future generations.

To maintain my connection with the real world, I wrote this book on trains, boats, planes and, with some difficulty, on buses. This brought me into some interesting conversations with comments like:

'You might change soil but you won't change people.'

'Ah you're one of those who think all men are your brothers, then?'

'You need to stop those immigrants from ripping us off and taking our jobs.'

'Do you really believe in global warming?'

'Why should we work when others can't be bothered to get out and find a job?'

'The economy has to grow for us to survive.'

'We have always managed to adapt to problems in the past.'

'OK, so what are you, personally, doing about it?'

The result is a book about the hidden life of the soil and about the hidden forces that drive you and me. Developing an awareness of hidden problems and opportunities, and acting on them by searching for beauty, offers some pointers for improving our lives. Beauty ultimately comes from our inner landscape and often appears when we are waiting on the spiritual. The searching itself brings awareness of our every action, our environment and our resources and has a natural outcome in a desire for the good of others.

Note:

1   The concept of deep ecology was attributed originally to Naess 1973 and an example of its development is given in Capra 1997.

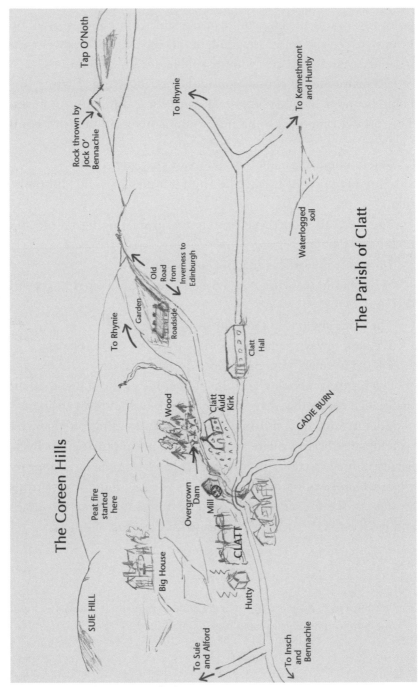

Figure 1: Map of the parish of Clatt

Chapter 1

MOLEHILLS AND POEMS

There it stands, the dear old house
Up on the grassy hill
It knows my joys and sorrows
And holds them secret still
Henry Ball

We all need somewhere decent to live, to put down roots, to belong. Thus I was overjoyed when my father announced that he had bought a house with a big garden in the village of Clatt. Two years of living in cramped conditions in a caravan stuck in the middle of nowhere would be coming to an end. I still remember the first time that we came to Roadside Croft. We took the bus to Kennethmont and then set off on the two-mile walk to Clatt (Figure 1). As we walked, I looked round every corner for the house. Eventually, it came into view, on its own on the horizon. Once we got to Clatt, I discovered that Roadside was not actually in the village. We had to walk another quarter of a mile uphill through the surrounding fields to reach the house. The track was lined with telegraph poles. This was truly exciting. Not only electricity, but also the phone! But while my father spoke about getting it connected throughout our time at Roadside, there was never any money, so the poles and wires remained mute, carrying only the chill hum of the winter wind.

The house had an interesting history. The land in the parish of Clatt originally belonged to two lairds. A laird is a member of the Scottish gentry with a heritable estate who usually owns most of the land round his large house or castle. Most of the land of Clatt was owned by the Knockespoch estate, with a small strip to the north owned by the Leithhall estate. The land on which Roadside was built was bought from the Leithhall estate by a souter, as a shoemaker was known. Previously, he had rented a croft from the Knockespoch estate but had been evicted by the laird due to misdemeanours involving a woman. Aggrieved, the souter bought a plot of land as close to Knockespoch as he could and built a house aligned so that it was clearly visible from the laird's 'big hoose'.

When we entered the house we were dismayed by the state of it. We began to understand why the Old Man had managed to get it for only £100. The previous owner, an aging spinster and a clergyman's daughter, had lived there alone with several dogs and goats and many fleas. We were soon to discover that we needed to be as hardy and persistent as those fleas to survive at Roadside.

The spinster died naturally, but her body had lain for several days before the clamour of the dogs trapped inside the house alerted the postman. In their hunger and desperation to escape, they had chewed the window frames, the doors and the furniture. The dogs were so desperate that the vet had to shoot them through the partly open front door before anyone could get into the house. As a result there was blood splattered over the doorposts and window frames. We bought the house just as the previous owner had left it, so the Old Man had to clean up this dried blood before we could settle in.

As we explored the house, we discovered small pre-war watering cans, solid clothes irons and of course the chewed pre-war furniture. Many of the items could have gone into a museum. We kept some of the furniture and the irons, but we had to get rid of most of it. There was no refuse collection lorry in those days, so we spent much of the first few weeks burying the effects of the deceased in a big hole dug in the garden. This was how we were to dispose of most of our rubbish that couldn't be reused or burnt. Today, we'd have been seen as environmentally friendly and acting sustainably. But we hated it at the time. Only in the last few years of living there did we carry rubbish to the bottom of the access track for council disposal.

The house wasn't quite what I, or more especially my mother, had expected (Figure 2). It had been two adjoining cottages and a sort of conversion had been made to bring them together. One cottage was used as the main living area and had three rooms, each having an attic above. The other had one main room which contained the water supply, the chemical toilet and an attic above. We called this the kitchen. However, nobody had linked the cottages together with an internal door, so to get to the kitchen or – more importantly – the toilet, you had to go outside the main house. This was inconvenient, especially in the freezing cold winters. Though we called it 'the kitchen', nothing was ever cooked or eaten in there and it was more like an outhouse.

I was nine years old when we moved in and Roadside Croft was full of mysteries for me. One great mystery was where the 'road' was in Roadside. The rough access track up to the house stopped at our front gate. The Old Man told me that the area beyond the gate where we were burying rubbish and starting to build sheds was part of an old road. Our land behind the house, or 'property' as the Old Man insisted on calling it, was in the shape of a triangle which extended to a point furthest from the house. The old road ran along one of the long sides of this triangle, was walled on both sides and was heavily overgrown. The local farmer told us that the road was the former main road between Clatt and Rhynie and that it had been superseded about a hundred years previously by two roads which ran round the hill on which Roadside was situated. Much later, digging through archives revealed that it had been part of the King's Highway between Edinburgh and Inverness. Making use of this road almost doubled our land area. We used it like our own even though it wasn't on the title deeds. It could not have been well metalled as we made several crops of hay from it. The ownership of this land continues to be debated to this day.

With precious little money and no running water, living at Roadside was a challenge. We lived mainly in one room where mother did the cooking and the washing up. In the early days she used a tiny paraffin stove. The house was heated by one fire in the living room and another in the main sitting room which doubled as my parents' bedroom. Between these was the chilly bedroom occupied by my sister and me.

The Old Man was unemployed so he worked in the garden, cobbled together bits of wood for sheds and looked after our hens. I loved it when he made hay, confidently swinging the scythe through the tall grass, full of clover. I loved the smell and the crinkly feeling of lying in it. Twice a week he would cycle three and a half miles to Rhynie for groceries and fresh bread. I enjoyed accompanying him on these trips when I eventually got a bike. In the afternoon, he would retreat upstairs to his study to write hymns, poetry and songs. This study, an attic bedroom, was strictly off limits and we had to keep quiet when he was there. When he was out, I would sometimes sneak up the stairs to peer at the masses of papers and books strewn around, with the typewriter in the middle. I think the Old Man felt that writing was his true calling – perhaps it was. He certainly loved Roadside with a passion. The isolated location with its rich spirit of place and the ever-present wind inspired him and enabled him to write with few interruptions.

As well as not being very fit, he had a problem in that he refused to do anything that he was told! Nevertheless he expected my mother to do as she was bidden. In the morning she did the housework and worked in the garden. In the afternoon, she usually hunted for sticks of wood for the fire, often going considerable distances, because we couldn't afford much coal. I often went with her and we would trail long branches up to the house for the Old Man to cut up. She had a rival in the village who also collected sticks and they did not always see eye-to-eye.

However, life wasn't all that bad. We bought our main weekly groceries from the large Co-op van which came from Rhynie. I was always amazed at the sheer variety of food and hardware it carried. I watched for it arriving at the bottom of the hill and happily scampered ahead of my mother to meet it. One day, when the Old Man was mending the track and we were trudging up the hill with the groceries, he got us to help him with some fencing. When our backs were turned, the cat got into one of the bags of groceries and seized a sausage. As the Old Man gave chase, the cat ran off with the sausage, but, of course, it was linked to the other half pound, so they bounced along behind him. I think I laughed till I cried, even though I had been deprived of my tea.

Although they had neither money nor jobs, my parents were both industrious. The Old Man used the scraps of wood and metal lying around Roadside to repair and build numerous sheds. Later, when I was 17 and bought a van, he even constructed a garage with a rough inspection pit. It was in the garden, though, that the Old Man excelled. Roadside was 240 metres above sea level and caught the wind from all directions. Nevertheless, the land was on fertile soil – which I later learned was of a type known as Insch series, made from rock rich in plant nutrients. A sign of this fertility was found in the many molehills in the garden and the surrounding fields. The soil bursting from the landscape below revealed its rich, golden-brown colour and fine crumb structure. My Old Man knew how to grow fruit and vegetables. I always remember the day when he parted the leaves of what looked like an ordinary cabbage and said: 'Look at this,' standing back to let me see the tender white flesh of a little cauliflower. 'Who said I'd never grow this up here?'

He taught me how to prepare the soil and, perhaps more importantly, how to reclaim the soil from rough grass. He also showed me the principles of rotation and the importance of legumes in the rotation. He knew what he was doing. He grew enough potatoes and vegetables to feed us almost every day of the year.

My first encounters with the spiritual were the usual negative ones that kids shared. This was exaggerated for me because the Old Man believed in ghosts and in the random direction of fate, usually negative – all of which made me rather fearful, as Roadside was dark and creaky at night.

As I grew up at Roadside I was constantly in contact with the land and the soil, so it was natural that I developed an affinity with the countryside and with the soil itself. I was first made aware of this at home, when I was about ten years old and somewhat bored in the summer holidays. I was sitting on the path alongside the cabbage patch at the top of our extended garden in the brilliant sunshine of early summer. Everything was still and quiet and the fields around the garden were silent. The walls surrounding the garden were low, broken down and made of the rounded granite boulders typical of the north-east of Scotland. These boulders attracted lichens, which could give them a crusty, golden, dappled appearance. I looked at one of these and then started to stare at it. I felt a growing sense of affinity with this stone, as if it and I were as one. It was a pleasant, warm feeling. I realised that, although this was an inanimate object, fate had arranged for me to feel a sense of community and linkage with the earth. I was sensing something below or within the stone. I rushed into the house to tell my mother. She was not amused and told me to go and find somebody to play with. It came as a shock that the person I was most close to could not understand something so important to me. I felt hurt and bewildered. I realised for the first time the difficulty and embarrassment that most of us have in attempting to communicate spiritual feelings to others.

I later realised that this spiritual experience came from the intensity of my connection to Nature through the stone walls and floors of the house, and through the plants, the air and the soil. There was also a connection to those who had gone before, those who had stood on the soil that I stood on and who were now enriching the spirit of the community as well as the subsoil in the graveyard.

As for the village itself, the word Clatt comes from the Gaelic word 'cleith', pronounced 'Cleit', meaning concealed from view on every side. An old poem, cited by Cuthbert Graham, says:

> *They call it Clatt – 'a place concealed',*
> *Yet Suie's Cairn its beauty has revealed.*

The Gadie wimples by its storied mill
And brave Knockespoch on the Bishop's Hill
Stands sentinel o'er copse and pond
And all the farmlands beyond.

The village is located at the centre of a triangle between Huntly, Alford and Insch and is surrounded by hills – the undulating Correen hills to the south, dark, flat-topped Tap O' Noth to the north-west and windswept Bennachie and Dunnydeer to the east. The minor road linking Clatt and Alford climbs the Correen hills to the best road view in Aberdeenshire. From here the parish of Clatt is spread out like a map with Knockespoch, the 'big hoose', separate from the little village with its school, kirk, and hall and the farms and the land supporting it round about. While the village of Clatt was small and we were on the outskirts, the parish covered an area of about 2,400 hectares with a population of about 350 and with Roadside near the centre. From the middle of the village, I could look up to the churchyard with the road rising in front, a stream and waterfall to the left, upward slopes on both sides and the kirk looking down from above. All soft curves, water and spirit – a good place to be. As kids it was the natural place for us to hang around and to play.

The villagers soon realised that we had little money and, if I appeared at the door unannounced, I was usually taken in 'for a piece' (a snack) and given 'something to tak' hame tae my mither'. Despite all this kindness, our family had a clear position in the pecking order of the village community. Naturally, the laird was at the top of the tree. This was well justified as he was caring, supportive and generous and consequently well liked. At the bottom of the tree was a family who, like us, lived off the National Assistance. We were not quite at the bottom, mainly because, I think, we were all considered to be mentally sound. Owning 'our own property', as my father boasted, didn't amount to much.

Status in the community was conferred by position, influence and willingness to participate and share. We were a little short on all three counts and I accepted our place in the pecking order without question. That was the way the community functioned. Money, or the lack of it, was less important in those days. Even though we depended on state handouts, our overall income and standard of living were not vastly different from that of the other villagers. Help came in the form of physical assistance, lifts or shared food and this all seemed to happen automatically. Some participated more than others, with a few, like the Old Man, resenting others 'knowing their business'. There were plenty of disagreements, but the mutual support gave a sense of security even in homes isolated from the village centre. The various community organisations were run cheerfully and usually without any feeling of obligation by those in charge. It was irritating not having much pocket money, living in a cold house and being without a car. But I loved the space and the wildness of the place and the freedom to explore and enjoy it. Also, being at the bottom of the tree had its advantages – I was close to the soil.

As I grew up I got fed up with tilling the garden and wanted to get on with 'men's work' with the soil. Most of my friends were farmers' sons so I grew up helping out on the farms. My main ambition at the age of about 12 was to drive a tractor. As kids, we spent hours talking about tractors, using terms like 'full throttle', 'into top gear' and 'she can fairly rug', meaning good torque and traction. I remember the first time I drove a tractor, an old Fordson Major. I had been asked to move the trailer when we were collecting turnips, or neeps as they're known locally. I climbed up, put it into gear, let up the clutch and thrust the hand throttle fully downward – hey! full throttle! Not fast though, as I had engaged first gear.

As an older youth, I spent most of my weekends and holidays working on farms, gaining an ever-increasing love of the land and the people who worked it. I did most jobs – cultivating, carting, tending animals and rolling silage. Some jobs, like painting doors and windows and cutting thistles, were remarkably boring and I was rubbish at them.

The one job that was not boring was 'scarifying neeps'. At the large seedling stage, these were hoed by hand to thin the crop and to remove the weeds. The neeps were planted in ridges about 0.5 m apart. At about the same time as hoeing, the weeds between the ridges were cleared using a tractor-operated scarifier. A pair of discs guided by cutaway rollers straddling the rows cut the soil and weeds away from the sides of the ridges. Nobody seemed to like this job as you had to drive slowly and the rollers needed adjustment from the tractor seat to follow any crop sown off centre. But I loved it. It was utterly absorbing and time seemed to stand still as I moved. The dark, moist soil carved away from the ridge leaving a sharp edge between it and the dry soil around the crop. I could never tire of it, the smell of fresh earth, cut vegetation and well-combusted diesel. And the sound of the gritty swoosh of the cleaved earth and the purring of the untaxed engine. Also, if it came on to rain, I was in the dry and the hoers would curse at me as they put on their oilskins: 'It's a'right for you sittin' on your backside a' day' and 'A few hours hoein' would sort you oot' and the inevitable 'You students dinnae ken whit real work is!'

It was a long time before I got to do really responsible jobs like sowing seed, combining crops or ploughing. Eventually, as a student, I started ploughing for a farmer from a nearby parish who had an alcohol problem. He was quite amiable when drunk, which was often quite early in the day. If he had had a few when he came to see me, he would appear driving his car flat out in first gear. Presumably, this was to keep his speed down and to minimise the danger to himself and to other road users. Things were a bit more relaxed in those days in country areas when car insurance was still seen as a bit of an optional extra. But the ploughing was good. I never tired of watching, smelling and hearing the endless ribbons of soil curving off the mouldboards. Ploughing links body and soil naturally even though machines are involved.

Ploughing also taught me that many kinds of soil could lie together in one field and gave a unique insight into the landscape below the surface (Figure 3). Different soil types produced different finishes. Sandy soils, ploughed fast, could give quite a level surface where you could hardly see the furrows. Heavier, wetter soils could not be ploughed so fast and produced more distinct furrows. Like most ploughmen, I aimed for beauty and quality, taking great pride in ploughing straight furrows of even depth. Different kinds of soil often made it hard to keep the furrows straight and even, and ploughing could be hazardous at times. One of the more remote farms contained some very steep slopes which could only be ploughed downhill. Travelling uphill with the plough carried on the lift resulted in the tractor's front wheels lifting off the ground and pawing the air.

Ploughing gave me about the greatest job satisfaction ever, as I could see my work for months afterwards from many viewpoints. I ploughed whenever I could. As this often coincided with Christmas holidays, I'd be out on Boxing Day and even New Year's day. As I left a field uniformly black at the end of the day and closed the gate, I sensed what Lewis Grassic Gibbon described as 'Clean and keen and wild and clear, the evening ploughed land's smell up in your nose and your mouth when you opened it.'

Like all jobs though, 'there's aye a something' to annoy you. Stones caused continual wear or breakages of the metal plough furrows. In some fields I soon learned where these were and to avoid them, an advantage of knowing the landscape below. Even so, I was constantly surprised at the size of the boulders the plough could bring up and often from areas ploughed several times before at the same depth. We were convinced that these grew out of the subsoil. Another irritation was the soil sticking to the plough furrows when working Insch series. I had to get off the tractor at regular intervals to scrape this off. The 'wonder soil' had a drawback. In my first soils research project at the University of Aberdeen I made a meter to measure how much the soil stuck to the metal. Like much research before and since, this helped quantify the problem but didn't supply a ready solution.

I took for granted the communication between the land and myself during ploughing. After a couple of years, the farmer bought a large, new tractor where the back of the cab was enclosed by windows and with mod cons like heating and a radio. This was great in that it allowed you to cover more acres per day in greater comfort, but it came at a price. I hated looking through a window at the soil. The window was often dirty and I felt remote from the soil, which I could no longer hear and smell. Vintage ploughing, where the ploughman is in the open air within easy reach of the furrow adjusters, allows fuller awareness of the land. But there are advantages to bigger and more comfortable tractors. One of these is that older and less fit farmers can carry on working. I have a farmer friend in Clatt who, at the time of writing, still ploughs all his land and is not far short of 80 years old! He has calculated that he covers 440 miles while ploughing the 300 acres of the family farm. This is a far cry from when he started ploughing with two horses and only managed to cover one acre per day.

When I left Clatt to take up my soils career, I continued to feel the closeness of the land. The spirit and the connection remain strong.

Ploughing

The plough swings round the tractor
and thuds on to the earth
A moment's relaxation then the land
grips the plough.
Screefers paring, bodies heaving
Hand on wheel, hand on hydraulic lift
Foot over brake, eye on furrow wall
Balancing sense, grip and draught
Waves of soil glisten in a curling ribbon
Lateral failure, loosening, inversion, burial

A thousand times over
Soil
exposed to weather, birds and man.
Bubbling through the breeze
the fresh smell of the living land
Uniformity created, fertility restored
And the stones tug at the plough bodies
like the land rugs at the heart of you
Sand and clay, dark and light
Coiling, never ending, yet finite
Ground of our being, our Mother earth.

screefer: blade that cuts the soil ahead of the main ploughshare
rugs: pulls

Figure 2. Roadside Croft and the Old Man
The car is sitting on the old road.

Chapter 2

HOLY GROUND

Observe, and let the soil tell you its story
Tom Batey

I learned a lot about the soil from the vast amount of it that I moved from an early age – digging holes to bury rubbish, sinking wells, and turning it over in the garden and in the field. This led me to want to get to know the soil even better, to understand it and find out what makes it tick. Most people get to learn about it from gardening or from seeing it on the television where it usually conceals either something like a dead body in a real or imaginary crime story or archaeological remains. They don't realise that the soil itself is the real buried treasure. I think that, in the mind's eye, soils are mainly seen as crumbs which are found in mole-heaps, garden seedbeds and lawn turf. Crumbs are indeed the building blocks of the structure of the soil.

Structure

If you take a handful of soil and look at it carefully, you're observing its structure: the size and shape of the crumbs and the lumps. Crumbs are clumps of soil particles held together mainly by the sticky residues from roots and microorganisms. These particles are formed by the weathering of rock. They are just like miniature versions of ordinary stones and come in a vast range of sizes. The biggest are sand particles, such as beach sand, which are typically around a millimetre in diameter. The mid-sized particles are called silt and are around 0.01 mm diameter and the smallest particles are clays which are very small, roughly 0.001 mm diameter or less. In between these there is a wide range of sizes. To get an idea of the scale of the difference, if a 2 mm ball bearing represents a clay particle, then a sand particle would be a sphere 3 metres in diameter. This would reach from the floor to the ceiling in your living room, half filling it.

Soils come in all shapes and sizes. They are often described in terms of how much sand, silt and clay they contain. This is the texture which can be estimated by rubbing wet soil between the fingers. Sandy soils feel gritty and don't stick together well, whereas clay soils can be moulded, a bit like plasticine, and the particles tend to stick together. Kneading a soil, especially a clay soil, is enjoyable and is not far removed from the experience of making ceramics or sculpture.

Living matter

The sand and silt particles stick together by means of both the clay and the organic matter consisting of plants, animals and tiny creatures, both dead and alive. The dead matter is bits of plants and the bodies of tiny soil organisms. This rots down to give a fine, dark, sticky material called humus which is chemically stable and gives the soil its dark colour. The living matter is mostly bacteria, fungi, mites, beetles and worms. The bacteria and fungi make up the soil microbes (microorganisms) which are so small that a handful of soil contains billions of them. All of these creatures work together so that the soil teems with what Stephen Lewandowski describes as wild life[1]. Root hairs thread through the spaces in the soil; bacteria and fungi decompose vegetable matter; bacteria hunt protozoa; slime moulds consume bacteria and fungi; soil fungi intertwine and live with roots and nematodes; and earthworms work down vegetable and mineral material. This mixture of the dead and the living suggests the natural intimacy of life and death. All work together in seamless co-operation, making the soil a wondrously complex but highly efficient factory.

The organic matter 'glue' is particularly important in soils without much clay, such as the Insch series where I grew up in Clatt. Organic matter also allows crumbs to stick together in bigger soil lumps of up to 20 cm diameter. Organic is a word that has a wide range of meanings, some of which seem far remote from a living organ:

Organic matter

Organic –
Teeming life
Bits of grass
Gobs o' shite
Gnarled roots
Black or white
Peat stores
Hydro carbons
Healthy foods
So we gather
Holistic thoughts
Or worthless chatter
Life-saving drugs
Living landscape
Lost forever
Growth and life
You and me
All together
– matter

I often wonder if there are parallels between the structure and life of soils and that of communities and families. Soils perform functions that are remarkably similar to those within a human body or community. Organic matter resembles our own energy and vitality. If these processes work in a healthy body they make us resilient to all kinds of stresses. Extending this analogy further, just as organic matter is what holds soils together, perhaps there is an invisible 'glue' that holds people together in families and communities, such as the shared activities of work, eating and drinking, rituals and religion.

If this glue weakens through the diminishing of any one of these activities – such as drinking at home rather than in the pub, or a decline in attending church or community gatherings like annual fairs – then our links to each other are undermined.

Soil degradation

Soil structure is really important because, unlike many other soil properties, it can change quickly in front of your eyes. Try standing on a molehill or throwing water from a bucket onto your freshly sown seedbed. These two actions change soil structure for the worse and cause soil degradation. The first action causes compaction and the second erosion, and both are common wherever human beings work the soil. Soils which are squashed together by compaction can end up as dense lumps, called clods, which need to be broken up to revitalise the soil.

The organic matter 'glue' is very important in preventing degradation of soil quality, because it helps to maintain good structure and healthy soil life. The combination of good structure and high organic matter content gives a resilience which enables the soil to resist stress. As mentioned previously, this has a parallel in human existence.

Porosity

Structure includes the size and shape of the pores, which are the holes and cracks in the soil. The ratio of pores to solid matter is defined as the porosity of the soil. The pores usually contain both air and water. When they are full of water, the soil is saturated, and when they are full of air, the soil is dry. On average, slightly less than half of our handful of soil consists of particles of rock, a quarter is water plus dissolved nutrients, another quarter is air and the wee bit left (around 5%) is the all-important organic matter. The pores are the 'holes in the ground' and are important because all of the action in the soil takes place inside them – the drainage and storage of water, the flow of air and water to roots, and the storage of carbon and nutrients. For plants to grow in the soil, roots and organisms need to absorb nutrients from the water in the soil.

We depend on the use of solar energy by plants in creating our fuel and food. The remains of ancient plants provide us with oil and coal. When we burn these, we are consuming solar energy that was stored millions of years ago. Max Purnell[2] has compared the soil to a rechargeable battery. The soil is regularly charged by solar energy through additions of organic matter, sugars and soil life. Nutrients are stored in the organic matter and the fine particles and are brought into solution by microbes. The microbes work in the pores, usually in isolated wedges of moisture. As a result, the different types of microbes are also isolated so that a wide variety of them work together in the soil at any given time. This helps the soil to have high microbial biodiversity, like a hidden rainforest under our feet. Some bugs change nutrients into a form suitable for crops, others change organic matter into humus, and others clean up pollutants and so on. If the bugs which release to plant roots the nutrients locked up in the organic matter of the soil were to suddenly stop working, trees and other plants would die out in a couple of years. Again this makes me think of human society. Our interests and our abilities for sustaining life are many and varied. Accepting people of different races, religions and cultures within our communities allows us to work together, each to his or her talents, to invigorate our society.

Despite all this activity, the soil doesn't grow much – or, if it does, it grows very slowly, the topsoil deepening under favourable conditions. The simultaneous operation of a complex set of mutually beneficial processes in the soil, based on recycling, provides a good model for a steady-state economy, which is something we should be aiming for as a society. Our politicians are constantly striving for economic growth to take us out of recession and maintain employment.

But economic growth is unsustainable because it involves overconsumption of natural resources, overproduction of wastes and degradation of the environment and society, all of which compromise national security. A steady-state economy, as proposed by Rob Dietz and Dan O'Neill[3], stays within ecological limits by aiming for stable or mildly fluctuating levels of population, energy use and material consumption, and leads towards justice, employment and equity for all.

Soil structure can be described either in terms of the shapes and sizes of the crumbs and lumps or in terms of the sizes and interconnections of the pores and cracks. I've spent a lot of my working life trying to measure soil structure using both methods. Soil structure can be seen as the architecture of the soil, with each soil crumb or lump representing a house. The sand and clay particles are like the stones and mortar of the house and the pores are like the corridors and rooms. The rooms are important because bacteria live there.

When the soil 'building' falls apart due to compaction or erosion, the loss is like a demolition. The value of the material making up the structure is lost because the rooms and corridors have been destroyed. Loss of porosity decreases the potential of the soil to grow crops and to do other jobs for us. To extend this idea to ourselves, a soil with good porosity and a wide range of interconnected pore sizes is like a person who is open-minded, who reaches out to understand where others are coming from and who fosters life and creativity. The house model can also be used as an analogy for larger ecosystems, describing the relationships between all the organisms living together in a particular area, and their relationship with the air and water and all other parts of the environment. Ecosystems range in size from a soil crumb through a forest to the whole Earth.

Describing soil porosity is useful but to start to really understand soil you need to look at it with your own eyes and experience it with your other senses, preferably where it is. Dig a hole and look inside or dig out the soil, break it up and then look at the bits. Smell it, feel it and rub it through your fingers. I find that the way the soil falls apart when lifted out by a spade tells me something about its structure and health. Soils are not easy to describe because one spadeful looks pretty much like any other. Tom Batey, my soils lecturer at the University of Aberdeen, used to say, 'Let the soil tell you its story.' However, this is not easy and we need help to know what to look for – much the same as we need to be trained to spot wildlife in the countryside.

Learning to look

I developed, with colleagues, a spade test[4] in which a slice of soil is dug out directly from the ground (Figure 4) and compared with a simple picture chart that shows the different types and qualities of soil structures (see Figure 5). I have demonstrated this at meetings with students, farmers and researchers where we usually compare slices of soil which people bring from their own fields and gardens. The slices are put onto tables. I usually demonstrate the test, then I get each person to pull apart their own soil and describe it to the others in the group. They estimate the quality of the structure from how easily it breaks up, from the shape and size of the lumps and crumbs and roots, and from comparison with the photos in the chart. I usually suggest that people take photographs of the soil slices as they dig them out. These can be compared side-by-side later on. Soils which look so alike when they are dug out can show big differences in colour and structure in photographs, especially if these are 'blown up' digitally.

Estimating the structural quality really gets people talking, especially the land users – those who actually work the soil. I'm always surprised at the sheer variety of soils that appear. Often farmers will bring in problem soils where crop growth is poor; sometimes these turn out to have good soil structure and the problem may lie in the nutrition of the soil. They share ideas on what the soil is, how it got to be in its current state and what, if anything, is needed to improve it. The test can be used for deciding the type of tillage or drainage required for good crop growth, though in real life farmers don't seem to use it much because they are often reluctant to get out into the field to dig holes. The chart is usually printed double-sided and then laminated in plastic to make it waterproof. My wife calls the A3-sized versions place mats. It's good to eat your dinner off them as you are constantly reminded of where your food comes from – and also the plate and knife and fork.

The test is frequently used in the field. I have colleagues in South America who use it because the structure of the red soils in central and southern Brazil often becomes hard and cloddy after many years of crop production. The soils work hard in Brazil because in many areas two crops are grown each year. When travelling abroad, I sometimes think that the method might not work well in unknown soils. But no, as soon as a spadeful comes out and is broken up, there is usually an immediate, shared interest in what the soil has to show. I'm pleasantly surprised at how often I discover something new when I use this method. The soil knows no boundaries between nations.

The act of digging up a spadeful of soil and gently pulling it apart is a positive, peaceful and even healing experience. This feeling depends on time and place but leads to a sense of the sacred – 'holy ground'. I like to handle the soil gently, to treat it with reverence, so that it will reveal more of its true character. One of the advantages of the spade test is that all you need is a garden spade and the chart. Seeing is believing. Pick up a spade and try it.

Looking deeper

I first saw what the soil was like in a deep hole when the Old Man started digging wells in his never-ending search for water at Roadside. These holes extended up to 4 metres deep. He was able to get to this depth without hitting hard rock because the soil was so deep and the rock of the parent material was quite soft, though I still don't know how he did it as he was not very fit. The rock of Insch series is olivine norite which breaks up into weathered shards, like the skins of an onion. To look at the soil from deep inside a hole is the complete soil experience. As I stood at the bottom of that well, looking up to the circle of sky above, smelling the soil, feeling its cool surfaces, I felt very much part of the soil and sensed its power. I also learned some of the language of the soil, the faint tearing sound and the soft thumps as slices of soil fell away from above my head and landed at my feet, warnings of that power.

Holes dug specifically to look at the soil, called soil pits, are usually about 1–1.5 metres deep and are long enough (3–4 metres) and wide enough (1 metre) for people to stand up in and to examine the soil (Figure 6). Yes, the way I describe it makes it sound a bit like digging a grave. The main difference is that it's wider and there are usually steps into it. The first time I looked down at soil in a pit I was surprised by how featureless and, well, boring it looked. I needed to get in, get up close and use a knife to gently pick at the surface to reveal the pores, the structure and the roots. I could then describe the soil by looking at the sides of the hole – the soil profile.

The most noticeable property in a profile is the sharp boundary between the topsoil, which is dark due to the organic matter, and the subsoil which is usually lighter. Describing a profile allows a really thorough investigation of the landscape below the surface. Although the description is similar to that in the spade test, much more soil is visible and changes can be seen in both vertical and horizontal directions. Looking vertically down allows identification of the natural layers and the man-made layers (e.g. compacted zones). Looking horizontally across allows detection of variations in soil structure and compaction caused by tillage machinery or wheels. Examination of roots, especially those below 1.5 m depth, often gives a good indicator of the overall health of the soil. Other things to look out for are organic material, earthworms and other visible bugs. This is a very thorough method, though its main disadvantage is that it takes up to 3 hours to dig and describe each pit. The spade test is much faster, taking only a few minutes. The pit method developed in France for diagnosis of cropping problems is called 'le profil cultural'[5]. This method involves the use of old-fashioned bellows, similar to the ones we used to get the fire going at Roadside, to blow away loose soil to show the structure.

We compared this method with several others at a scientific meeting in 2005 at Péronne in France. Péronne is close to the River Somme and was badly damaged in the First World War. This meeting proved to be rather poignant as we were describing a good quality, deep loamy soil which our forefathers had dug as trenches and fortifications. This time, as we sank below the surface in soil pits, we were not using the soil to conceal ourselves, but were trying to reveal what the soil had to show to us.

Soil creation takes a very long time. It takes 10,000 years or more for ice, rain, wind, sun, vegetation and bugs to make 30 cm of soil. So when you dig a pit you pass through thousands of years in a couple of metres of soil. As you look down the sides of the pit, you can imagine how the soil was made – the movement of the ice, the scraping of the rock surface, the particles floating down through the water, the roots spreading downwards. You're looking through time as well as through space. The other advantage of being in a pit is that you become aware of being held within the biosphere rather than standing on the surface of a planet. L. Sewall believes that this produces a feeling of vulnerability and liberation[6]. Also a feeling of perception which is like a communion with the soil and which she believes can be experienced as a spiritual practice.

Farmers are often willing to dig pits using mechanical diggers as it's a lot easier than using a spade. A disadvantage of this is that they often leave no steps for you to get into the pit, so sometimes you can lose your footing and you get in quicker than expected:

A step in time

Ouch
I've just slipped across
ten thousand years of time
in soil.
My foot left the dark of the present
to the light of the past.
Carbon free – almost.
Soils of times under my aching sole
a PhD in a pit.
Doctor Who?

Variety

There is a lot of soil variation within a field, even over distances of only a few metres. I was recently involved in a long-term field experiment on organic farming in Morayshire. We had 28 plots, each measuring 23 metres by 40 metres, in a field roughly the size of four football pitches. We had noticed over the years that the crop yield from each plot depended on where it was in the field. I was given the job of finding out why yield varied so much with location. To do this I needed to look at the soils, so we dug two pits in every plot, making a total of 54 pits. You can imagine how popular I was with the guys who were helping me.

It was a typical sandy soil with no clearly visible structure. According to the local soil map the topsoil was about 30 cm deep and the subsoil was yellow-brown. I thought that I would find similar soils in every pit. Wrong. After looking at only six pits, the colour of the subsoil had gone from light buff through yellow to orange to brown – all in adjacent plots. Also the depth of the topsoil varied from about 27 cm to nearly 60 cm. The result of the investigation was that crop yields were greater where topsoils were deeper and less acid. This amazing variation put me in mind of how much people vary in character and depth of character when we probe below the surface. Perhaps it's little wonder that both crop yields and what we each achieve in life vary so much.

On a world scale, soils come in a huge range of colours from white to black, including purples, reds and yellows. Hildegard of Bingen, a German nun who lived in the 11th century, described the characters of soils according to their colours. She is considered to be the first writing soil scientist in Europe[7]. This was in addition to all of her other creative talents, which included composing music, writing books and poetry, healing, philosophy and seeing visions. She first became aware of having visions at the age of five and explained that she saw all things in the light of God through the five senses of sight, touch, taste, smell and hearing. A remarkable woman. She stated that a red soil was best as it had the right mixture of moisture and dryness and so produced a lot of fruit. A black soil also had a good mixture of moisture and dryness but was cold and was less productive. A green soil (I haven't seen many of these) was worst because it was both cold and dry.

The soil is best studied in its natural habitat and with minimum disturbance, as an integral part of our landscape. One of the disadvantages of digging a full-size pit is that the vegetation around it is trampled and a lot of soil is disturbed. The impact of soil disturbance was brought home to me in 2009. During my research, I have worked with colleagues on soils in field experiments at the Bush Estate, near Edinburgh, for about 40 years. There is one field in particular where we ran a sequence of experiments and gathered masses of soil and crop data which have been published widely. I decided to show this field to an international group of soil scientists who were visiting us for a meeting on soil structure. It was close to an area where a large-scale water purification plant was being constructed and I knew that earthworks were imminent. I looked at the field on Monday evening to check that it was OK for my visitors to see later that week. When I took them there on Wednesday, the field had disappeared. It had been dug up and hidden under a 15-metre depth of topsoil dumped there from another field. The soil profile – the soil's signature and identity – was gone. Not only had the field been violated, but I felt that a part of me and the integrity of my work and that of my colleagues over the years had also been violated. A precious, documented resource had been wiped out in a few hours by earth-moving (earth-destroying?) machinery.

People are perhaps unaware of such massive changes to soil and their long-term consequences, especially as they happen so quickly. Although soils can be lost in an afternoon during construction projects, soil creation and restoration are very much slower processes. Soil is degraded as more and more land is covered by buildings and roads, disabled when it is buried under half a metre of tar or concrete. This brings to mind the many people who have had their lives suppressed, smothered and trampled over by bullies or by oppressive regimes. The damage can happen quickly but leaves an indelible mark on a person's life. Restoration can take a long time and may never be complete.

I believe that by studying the earth and getting to know it better we develop a personal relationship with it. The poet William Blake spoke of being able to 'see a World in a grain of sand'. When we really observe the soil, we begin to love it and then to treat it with respect and reverence. The theologian Leonardo Boff summarised this well, saying, 'When you are studying the soil the cosmos is studying itself.'[8]

Chapter 3

WATER OF LIFE

… every drop of water in me (and you) has been in every
spring, stream, river, lake and ocean in the world during our
earth's billion years of existence. We are related to every other
self in the Universe.
Walter Wink[1]

The number one reason for my parents' falling out was water. Usually the well had run dry and somebody had to go and fetch water
in buckets. So, from an early age, I was made aware of the dominant
importance of water for family life. As well as the unreliable water
supply at Roadside, we had no plumbing, so that both the getting
of the water and the disposing of it were difficult and frequently
unpleasant. The cycle of water use was clear to us all. This made
me think that it would be interesting to follow the fate of some drops
of water through Roadside in order to show how important it really
is to our daily existence.

The water arrived as rain or snow, mostly coming in on a westerly
airstream. If it landed at the top end of the garden, on the old road,
then the drops hit the blades of grass and slid gently to the ground
where they sank through the well-connected macropores to be
stored in the soil. Eventually, after more drops joined from above,
the water slipped sideways and downwards into a network of tiny
underground streams which led into a larger underground stream
that snaked down the garden through the subsoil until it dropped
into the main well.

Our main well was on the edge of the croft in a small piece of
land just past the kitchen. Tiny though it was, the laird of Knockespoch estate attempted to secure it for a neighbouring tenant shortly
after we arrived. After some hassle with lawyers whom we could ill

afford, it did prove to be in our title deeds. In former times the well served the manse (church minister's residence) at the bottom of the hill below our house. The manse had long since been connected to the mains supply and had been taken over by a farmer who would occasionally tap into our supply for his cattle when they grazed the fields around Roadside. Either this or the dry summer weather would cause the well to dry up most years. It wasn't so much running water as water that ran away.

The drops of water were drawn from the well by a pump in the kitchen which would rasp and cough as you cranked the handle back and forth. When the handle suddenly became difficult to pull, the precious, clear, cool liquid appeared. Ironically, when the pump ran dry we had to resort to carrying water up from the manse, 400 metres distant. The situation could become fraught when the Old Man fell out with the farmer supplying the water.

The Old Man was convinced that there was more water to be had under Roadside so he decided to dig another well about a year after we moved there. To find water he used a pair of copper divining rods. Because they responded differently to different people, we all had a go at using them. They were L-shaped and you held the shorter part of each L in your clenched fists so that both rods pointed straight ahead and in parallel. You walked forwards holding these tightly and when they moved inwards and crossed, you were supposed to be above flowing water. When I used them, I was never convinced that it wasn't a subconscious effect that caused my hands to slacken a little and let the rods cross.

The first well that the Old Man sank yielded little water so he hired a professional diviner who located two more places. The diviner was sceptical about finding much water at Roadside because the catchment area behind the house was small. The second well he sank was no more productive than the first, but the third was best, though it proved no more reliable than the main well. The Old Man dug these wells deep. He lined the third well with concrete piping.

Somehow he managed to maneouvre these heavy pipes into place by hand. I still don't know how he did it. My mother was an unwilling party to his hare-brained schemes, often lugging out many bucketfuls of sand during these risky excavations.

Returning to the drops of water travelling through Roadside, they were carried in a full plastic bucket from the pump to the living room, were poured into a kettle and boiled for tea. After the water refreshed us we excreted it into the chemical lavatory where it sat for a week absorbing nutrients from the surrounding faeces along with, occasionally, some unwanted bacteria and viruses. Then the drops, now enriched with carbon, and dark and viscous, were buried in the garden, returning to the soil. The soil microbes thrive on such liquid, transforming the nutrients into a form that can be used by plants and also purifying the drops, restoring their beauty.

The Old Man buried all our wastes in various parts of the garden and over the years we would grow healthy vegetables on the soil. We trusted the natural filtration ability of the soil without question. The churchyard of Clatt is on a hill overlooking the main village and the Gadie burn and contains the remains of almost one thousand people (Figure 7). We take for granted that the soil of the churchyard will purify the rainwater going down from the churchyard to the village. However, in other situations there are limits to this purification capacity. A flyer which came through my door recently from the Water Aid charity stated that every seventeen seconds a child in the developing world dies from water-related diseases. In the time that it takes you to read this page, three more will have been added to the death toll. This adds up to a bewildering total of roughly two million infant deaths each year in rural Asia and sub-Saharan Africa. Every day, people in the world's poorest countries face the dilemma of having to drink dirty, contaminated water that could kill them.

My mother complained regularly about living with two children without running water. Our water was more like walking water than running water because so much of it arrived by foot. I certainly came to value every drop. We rued the day that one of us contaminated it by dropping something into the pails. We consumed about 5–7 litres of water per person per day. This is slightly greater than current use by the poor in arid regions of the world where women have to walk several kilometres each day to supply their families at a rate close to the biological minimum of 2–5 litres per day. Current water consumption in Europe and North America is boosted by the use of washing machines, daily showering, dishwashers and lawn sprinklers to reach a phenomenal 1,000 litres per day per person. We had none of these conveniences at Roadside and seldom had enough water to spare for the garden when it was dry, though we used waste water to keep the bugs off the vegetables. We couldn't even harvest rainwater as there were no gutters. We always knew when it was raining because the drips would start to run off the corrugated iron roof.

If the drops of water we are following fell close to the house during a spring storm then they might have struck the bare soil of a freshly sown seedbed. The big drops, landing on the soil crumbs like little bombs, would break them up into particles, some of which would be suspended in the drops. Some of the drops flowed into the soil through the macropores, which might eventually block up as the suspended particles were filtered out. If this happened the drops couldn't get into the soil and ran off across the surface to head downwards towards the house, gathering speed and more soil particles as they went. The drops soon encountered a 0.5-metre-wide drain laid on the soil surface and cemented to the back wall of the house to dispose of such run-off and the water dripping off the roof. Below this drain, the soil was in direct contact with the bottom 0.5–1.5 metres of the rear wall of the house. Due to the persistent rain, the soil below this drain was saturated and the drops flowed through it, penetrating cracks in the wall, seeping under the linoleum and emerging onto the floor of our main living area. Once, my mother noticed this happening at 11 pm and, complaining bitterly, got the

Old Man out of bed and they mopped up the water and the suspended soil particles into the night.

Soil is important for helping to control flooding during heavy rainfall. In recent years, prolonged rainfall in south-west England caused widespread flooding and erosion and led to extensive soil loss and much damage to houses and business premises. If the average soil porosity of the land subject to flooding had been only 10% greater, the soil would have been able to catch and store enough of the rainwater to have prevented much of the flooding and runoff. Soil washed away during flooding of houses causes extensive damage as everything is left covered in a layer of silt when the flood water recedes. As well as the loss of the soil particles, the eroded material contains carbon and plant nutrients – useful, important components of the soil which have simply become pollutants.

I sometimes think that soil erosion by rain has a parallel in the constant bickering and arguments in some families which lead to their falling out and fragmenting, often with one or more people ending up on the street. Or, looking at it slightly differently, it could be an analogy for our constant bombardment by the relentless overload of information, advertisements and technology that causes us to lose our sense of what is really important in life.

Soil erosion is controlled by covering the soil surface with organic crop residues and manure or by growing plants. This protects the soil aggregates and increases the soil porosity, which helps the soil to absorb heavy rain and maintain its vitality. In comparing the restoration of eroded soils with healing, Elan Shapiro states that 'our souls cry out for a rich inner life and for a grounded, diverse community to slow up the bruising pace of our lives, to create a holding environment in which we can turn our trials into sources of strength and integration.'[2]

Returning to our living room, the leakage was slow so there was never any bad flooding, but my mother hated the prospect of water coming into the house and would threaten to leave. She, of all of us, seemed to suffer most with the water problems – from getting it, through using it, to disposing of the waste. In African countries such as Kenya where water is scarce, it is almost always the women's job to fetch it. Women have a central role in providing and using water and safeguarding the environment, yet they are seldom empowered to have a role in equipping or taking part in water use and conservation schemes.

Scotland is famous for its freshwater lochs and its salmon rivers, which are extensive and plentiful. Yet, surprisingly, more water is held in the soils of Scotland than in all of these freshwater sources put together. Worldwide, 70% of the available fresh water, called green water, is held in the soil and is accessible to plants. To store and release this water to plants, good soil structure and porosity are needed. At Roadside, the Old Man was justifiably proud of his ability to grow potatoes. A potato crop growing fast on a hot summer's day acts like a big water pump, drawing water from the soil through the roots, up the stems and out through the leaves (Figure 8). In this way an actively growing crop moves vast amounts of water from the soil into the atmosphere, typically pumping up to 10 litres per square metre daily, equivalent to 100 tonnes of water per hectare every day. Put another way, for every tonne of dry crop harvested, between 300 and 600 tonnes of water are required to produce it.

The Old Man would water the plants in the spring and summer if the well was still running, a simple form of irrigation. Irrigation can make the desert bloom and is hugely important for food production, allowing farming in arid regions and offsetting drought in other areas. Irrigation produces more than one third of the world's harvest and 17% of the world's croplands depend on it. Some countries – such as Egypt, Sudan, Iraq, Jordan and Israel – are vitally dependent on irrigation for crop production. However, there are problems. From ancient times, short-term gains in crop yields resulting from the introduction of irrigation can often lead to long-term losses in productivity due to dwindling water supplies or to soil waterlogging and salinisation. The soil has a kidney function similar to our own, filtering out salts in waste water, but the process needs fresh water to flow down through the soil regularly and this may not happen. Data for 2007 showed that soil salinity was excessively high over an area of about 1 billion hectares. This reminds me of how we can boost human productivity by working long hours or by taking stimulants or drugs but it proves unsustainable in the long term as we become sick.

On our televisions we regularly see starving people in countries like Ethiopia and the Sudan where drought has caused crop or pasture failures. But the problem is more than drought caused by a simple lack of rainfall. It is the result of long-term land degradation caused by overgrazing and over-tillage, which reduce the ability of the soil to store water. Semi-arid lands occupy about 40% of the earth's surface and support two billion rural people, many of them poor. Land degradation makes the soil prone to wind and water erosion. Wind erosion is producing large dust bowls in China, Mongolia, Africa and the Middle-East[3]. These ultimately result in desertification that endangers livelihoods because the land can no longer support livestock and crops or provide wood for fuel. A recent example is the severe degradation of soil in Madagascar by overgrazing, burning and removal of forest vegetation. This has resulted in the loss of wooded land, erosion and thinning of the soil. The landscape (Figure 9) shows soil with scars and pockmarks on the surface skin and the rivers run red with the lost earth.

Already 70% of total water withdrawals are used for agriculture, yet global water demand is set to increase by 30% by 2030. The limits of irrigated farming are within sight. The increase of irrigated areas has dropped dramatically in the past decade as aquifers and rivers become depleted. Lester Brown reckons that we are close to 'peak water', a situation that arguably poses a threat greater than that of peak oil[3]. The solution needs to include better use of rainwater by reducing runoff and by increasing storage in the soil and in local dams. Crucially important for this is the organic matter in the soil. Organic matter increases the number of medium-sized pores which capture and store water. For example, increasing organic matter content from 1% to 3% can double the soil's storage capacity of water available to plants. We need to save and use every drop of rain, otherwise the drops lost will come to represent the tears of lost children.

Back with our droplets of water at Roadside, some escaped past the main well and ran down through the fields to the Gadie burn. On their way they may have picked up a few nitrate ions dissolved in the soil water shortly after nitrogen fertiliser had been spread over the field. As they went with the flow of the Gadie, these drops eventually lingered in slower-moving areas near the stream banks where the nitrate fed algae, resulting in a green bloom visible below the surface. Algae use up the oxygen in the water, a process known as eutrophication, so that fish and other freshwater creatures suffocate and perish.

Other drops, the very lucky ones, may have encountered that supernatural being that haunts some of the burns and rivers of Scotland, the water kelpie. This is a shape-changing creature from Celtic folklore which can at will transform from the enticing shape of a beautiful maiden back into a strong, powerful, breathtaking horse. The water kelpie is said to be malevolent, luring humans, especially children, into the water to drown them and eat them.

Most of the time, though, the Gadie is a cheery, peaceful stream that tumbles over a delightful waterfall in the village. It also contains wider areas which form slow-moving, deep pools. In the summer the sun casts dappled shadows through the leaves on to the bed of the stream. The midges dance above the water and the young trout dart along close to the bottom, disturbing the layer of silt as they go, producing little cloudy puffs. This landscape below the water surface is beautiful, peaceful and restful and more captivating of the soul than any water kelpie.

Water emerging from the soil tells of new life. In the field where we kept our caravan before we moved to Clatt, we collected our water from a spring located on a steep slope at the far end. Here the water trickled down from the peat above into a little pool before disappearing underground. In the sun this water glistened and sparkled and tasted of heaven. It is no wonder that springs and wells are often referred to as sources of the water of life. In many spiritual traditions, water represents the flow of life. Baptism with water is symbolic of rebirth into the family of God and the receiving of the Holy Spirit. Water is used to symbolise the cleansing of sin in most religions. Hindus bathe the bodies of those who have died with sacred water from the Ganges before cremation. The Maori people strongly associate water with spirituality. They permanently block recreational access to lakes where graves of their people lie beneath the water-line and also, temporarily, after an accident involving loss of life. They respect the role of the soil in the water cycle, stating that the purification of water by the soil restores its power.

Within us all is a spring of love that is activated and wells up when we work with the land or attend to the needs of others. It is a connection with the nature of God. Jesus, we are told, claimed to be the water of life. Those who drank from Him would never be thirsty, implying that the Holy Spirit would enter them and transform them for ever, perhaps by reinvigorating the spring of love. I believe that the activity of this spring needs to be stimulated in some way, religious or otherwise, in order to release more of the power of love and good from within us.

Chapter 4

GASES AND SPIRITS

I love the contours of the hills
The wide valley down below
The birds that wheel high in the sky
River's spate or gentle flow

And then there is the meadow land
With streams like silver threads
While all around the cattle graze
On damp turf where water spreads
Henry Ball

Air quality was good at Roadside. We lived there at a time before farmers sprayed slurry or pesticides on their crops. The only strong smells were those from muck-spreading in the surrounding fields and when the Old Man burnt rubbish in the garden, and these were not unpleasant. One of the best times to appreciate the good air was in the glorious gloamings of early summer, when the sky was cloudless and the lights from the surrounding farmhouses pricked into your consciousness one by one. I remember one night in particular when it was so still that I could hear a dog barking from the distant hill and the clickety-clack of a far-off train rising and fading as it threaded its way through the landscape. These sounds were carried through an atmosphere that made distant objects seem very close. Just above the surface of the land a faint blueish vapour hovered. Below this atmosphere the landscape stirred little as night fell. No incessant hum of distant traffic in those days. The Old Man loved to be out on such evenings and, as we stood together not saying much, for once his guard dropped and he was himself, fully absorbed in the present moment. It was then that the sensitive, understanding father who wrote poems emerged and we loved the landscape together.

Yet, for all the beauty before us, the soil was emitting greenhouse gases. We hadn't heard about global warming in those days, but it was already happening – the steady rise of global temperatures as greenhouse gases trapped the heat of the sun. Today, many are still sceptical about the big contribution that humanity has made to global changes in weather and climate, but their views aren't supported by the evidence of good science. This reminds me of the argument that I often had with the Old Man about coal. I used to try to persuade him that coal was made from wood, but he would have none of it.

'Coal comes from old leaves and wood,' I'd say, 'buried in the ground and compressed over millions of years.' I was proud of my scientific knowledge.

He'd counter, 'That's rubbish, it's dug out from among other rocks way deep down. And, look at it, it's as hard as rock.'

'No, it's carbon from years ago. Our teacher says that they can even tell the dates when the coal started to form. It's in all the textbooks.'

'No, don't you believe all you read in books.'

'But look at it, you can see the layers where the wood was.'

'No, I'm having none of your fanciful ideas.' He'd end the conversation by walking away.

On that evening, I was unaware of the greenhouse gases escaping from the land all around me. It had been a hot day and the soil was still warm, so the microbes near the soil surface would have been active in producing these gases. Like us they would have been breathing, absorbing oxygen from the air and releasing gas enriched with carbon dioxide. As water is the life-blood of the earth, so air is its breath. Just hidden from view at the bottom of the hill was a low-lying area in the corner of a field where water lay in the winter. In the springtime I would sail my tiny toy boat across it, letting it be driven by the daytime breeze. I was unaware then that when the water disappeared later in the spring to leave wet soil on which nitrogen fertiliser was spread, other soil bugs got to work producing an even more powerful greenhouse gas, nitrous oxide. This (also known as the 'laughing gas' used by anaesthetists and dentists) has 300 times the global warming effect of carbon dioxide. It is a very powerful greenhouse gas and soils are its main source, especially when they have just been fertilised.

The soil was not only wet but had an unusual grey-blue colour (similar to that shown in Figure 10) and a faint yucky smell. I was to find out later that this was caused by waterlogging, which is when the oxygen from the atmosphere cannot get through the soil to allow the micro-organisms to breathe, causing some to suffocate and others to start to produce greenhouse gases. As an adult, I sometimes wonder if this condition of the soil is not unlike ourselves when we get depressed. Life becomes too much for us and we start to suffocate under the pressure of work or family commitments, leaving us feeling blue and grey and not motivated to do anything much.

The cattle whose plaintive, rasping cries came to me through the darkening air were emitting methane from their throats, produced in the cud that they had been chewing. Methane is another important greenhouse gas, but is not quite as powerful as nitrous oxide. Every gram bellowed out by the beasts is equivalent to 24 grams of carbon dioxide. Their faeces also provide a ready food source for the bugs in the soil which produce methane from them and nitrous oxide from the nitrogen-rich urine. The grass and the cereal supplements used to feed the lowing beasts were likely grown with a nitrogen fertiliser. Oil is burnt to produce this fertiliser, resulting in yet more greenhouse gas emissions. Keeping livestock produces large quantities of greenhouse gases. Today, more and more people want to eat meat, especially in China and India. This means an increasing area of land is being used not only for grazing but also in producing cereal for concentrates used to make animal feed. Far from home, this leads to deforestation, particularly where tropical forest is converted to arable agricultural land, such as for soya beans, with huge losses of carbon (as carbon dioxide) to the atmosphere. The three greenhouse gases – nitrous oxide, methane and carbon dioxide – have a great impact because they remain in the atmosphere for a long time. Soils are particularly important as producers of nitrous oxide, contributing to up to half of the global warming effect of agriculture, which itself provides about 30% of world greenhouse gas emissions. On that night in Clatt, the clear, balmy atmosphere contained hidden threats. Nevertheless, things were not so bad there, as cattle are one of the few species of creatures capable of converting grass to protein and the ones I saw were grown with few concentrates in their diet. Grazed grass is also good at cycling and storing carbon.

I didn't know any of this then as I breathed in the cool evening air sweetened by the scent of the honeysuckle lighting up the garden fence. The calm was pierced by the 'coo-wee coo-wee' cry of a curlew as the air resisted the downward force of its beating wings, allowing the bird to rise. Tranquillity returned, intensified.

Earlier that day during the afternoon heat I had been playing in the wood that surrounded the clogged-up millpond of Clatt. I loved it there because it was wild and there were special, secret areas that you could reach only if you knew the right paths. There was a grassy clearing where I would lie down in the heat of the day and cool off. It was dark and restful on the eyes. I could look up through layers of branches and leaves and see the tops of the trees outlining a rough circle of deep blue sky. Each tree had a vast leaf surface area. I later found out that this was equivalent to a total area of up to one hectare or two times the size of a football pitch. The leaves of a rainforest tree have an even bigger area, up to ten times greater. The trees were breathing in carbon dioxide and using sunlight to convert it into oxygen and more leaves. No wonder the air felt good in the wood. But at Roadside that evening, the vast expanse of green crops, trees and grass surrounding me had stopped producing oxygen for the night and all of us, including the soil, were now taking in the gift of oxygen from these plants and were returning carbon dioxide. I was distracted briefly by the sound of a motor car driving towards the wood. Like the Old Man and me it was breathing in oxygen from the plants and producing carbon dioxide, but a lot more than either of us. The carbon dioxide from the car would lie around the wood until morning when the daily cycle of carbon gas exchange would begin again and most of it would be mopped up by the trees, crops and grass.

I thought of the yearly carbon cycle as well. The trees would lose their leaves in the autumn and the old branches would die off. These would either rot down in the earth or find their way into fires like ours in the house, releasing the carbon from the rotting material and from the ashes scattered over the ground. The carbon would be stored in leaves, fruits, branches and flowers. Beauty would be restored. It is important that some of the rotted material remains trapped in the earth as carbon; this is carbon sequestration, a significant way for soils to capture carbon dioxide from the atmosphere and help to combat climate change.

The farmers also participated in the yearly carbon cycle. They sowed seeds in the spring, and as the crop grew, carbon was taken from the atmosphere and stored in leaves, stems and grain and in the soil as roots. The crop was harvested and the grain used to feed us, either directly as porridge or through cattle as meat and milk or through fermentation and distillation as whisky. The straw was baled and used as bedding or feed for cattle.

I remembered when we had gone to an evening service at Clatt church the previous year for a harvest thanksgiving service. After the service, I emerged from the front door of the church into the dark with that faint glow of self-righteousness at having done the right thing. As I walked with several others in the darkness round to the back of the church we became aware of a deep red glow from behind the hill where Roadside was. There were no streetlights in those days, so any light was soon spotted and could be dramatic. The glow grew brighter and I could just distinguish yellow sparks in the wind. We saw the fire engine hurrying along the road linking Clatt and Rhynie. One of the farms was on fire. For a short time we watched, standing side by side, and I sensed the smell of burning in the air. The light was bright enough to make our faces appear faintly red. It was a cruel contrast as the slight ache in the throat caused by singing our thanks was replaced by the taste of burning and loss. Nobody said anything for some time and then some left to see if they could help, while the rest of us went home, knowing there was little that we could do.

Next day, it turned out that a barn of hay and straw had gone on fire. Hay ferments if it is baled when wet and can heat up enough to start burning. No life was lost and the farmer claimed against insurance, but the carbon cycle was broken and an excess of carbon entered the atmosphere. A whole summer's worth of work was lost as well, of course.

On another occasion, back in the 1930s, a fire started on the hill next to the Suie. These hills carry a layer of peat below the heather which usually remains sufficiently moist not to catch fire when the heather is burned off once every few years to encourage new growth. However, it ignited that year because of the very dry spring weather. The farmers fought the fire valiantly and initially managed to control the blaze, but it restarted almost one week later and this time it spread towards the big house at Knockespoch, located near the foot of the Suie hill. The laird ordered all of the tenants to beat and put water on the flames and dig trenches to control them. It is said that the fire was spectacular, the dusky red shining through the smoke as it grew dark. They worked from dawn till late into the night, coming close to exhaustion. Other creatures were also getting tired and very frightened. The hares, native to the hill and still in their white winter coats, were driven ahead of the flames. When confronted by the farmers, they eventually turned and ran back through the fire, stark white against the smoking, charcoal black of the peat, singed but unharmed.

As the fire burned and spread, several families in homes near the hill had to be evacuated. But, again, the greatest and most invisible damage was the immense loss of carbon to the atmosphere. The thicker patches of peat caught fire to depth and had to be dug up and spread out to stop them burning, a hazardous task. One third of the UK's 4.5 billion tons of soil carbon is stored in the peaty soils of Scotland and a large fire can upset the fragile carbon balance. The fire was eventually brought under control, mainly as a result of a drop in the wind. Eventually the vegetation re-established itself on the hills and once again carbon dioxide from the atmosphere was stored in the heather and bracken which then began the slow process of rotting down to form new peat.

Back in that darkening twilight, towards the bottom of the field and close to the village hall, I noticed a tractor sitting in a field. In the fading light I realised that this was a 'wee grey Fergie'. Just about every farmer had one in those days. The farmers had told me that this machine came almost as a direct replacement for the horse. Its huge advantage over earlier tractors was that it was compact and highly manoeuvrable, and had a hydraulic lift, so that it was almost as mobile and agile as a horse. The advent of the tractor was hastened in Clatt and surrounding districts by a disease which disabled some of the farm horses. The other great advantage was that the grey Fergie was light on its feet. Its rear wheel could run over your foot and, though it would ache, it was unlikely that anything would be broken. I wouldn't fancy one of the new, big tractors running over my foot. These can exert pressures of over 30 pounds per square inch. Imagine what that does to soil, especially when it is loose or wet and at its most vulnerable. The structure is squeezed out of the soil and it becomes solid, hard and grey (Figure 10). Yet today it happens all the time, and in a field near you.

Part of our desire for progress and for economic growth involves trying to get the work done more quickly, so we end up acquiring more and bigger implements. This is certainly true of tractors, which are getting larger and more technically efficient. However, as tractor sizes go up, so does damage to the soil because of increased compaction. Sometimes parts of fields are cut to pieces by deep wheel ruts. But the soil can be compacted deep down in the soil profile where it is unnoticed by the farmer. Soils suffer from violence in much the same way as we do. Running large tractors over the soil applies a lot of force over a short time and so is an act of violence of a kind. Compaction compresses the soil so that the porous airways block up and prevent the soil from breathing properly. This results in crop roots having difficulty in getting through the soil. Such effects of compaction are largely invisible, so it is widely known as the silent thief, robbing the farmer of his yield and the soil of its nutrients and structure. It has been estimated in a report in the *National Geographic* magazine that compaction costs Mid-West USA farmers $100 million in lost revenue every year.

Air is the source of life in soil and in us and acts as a life force within. The concept of Spirit is derived from the Greek word meaning breath and, like breath, is the power of life. Leonardo Boff[1] believes that Spirit was present right from the beginning of the universe, permeating it and emerging in a series of forms until it reaches its highest expression as the divine Spirit. Most spiritual practices involve deep breathing to create calmness and awareness of the present.

Later, when in New Zealand, I discovered that the Maori are a very spiritual people and are aware of this. Their traditional greeting between two individuals is the hongi, which involves the sharing of breath. They grasp right hands, put their left hands on each other's right shoulder and press foreheads together. The breath of life is exchanged and intermingled between host and visitor, making the visitor at one with the Maori. Their awareness of the presence of Spirit includes the soil. They consider the soil as one of their primal parents, the Earth Mother, and believe that she can release spirits. They are especially aware of this near Rotorua where geothermal activity causes steam to rise out of the land.

One of my friends at Clatt learned about the release of Spirit from the soil unwittingly. When we were youths at Clatt, in the winter evenings we got together as a Young Men's Club in the Petrie Clubhouse, commonly known as the hutty. It was donated in the 1920s by the widow of the owner of the general merchant's shop 'for the use of the young men of the parish to meet on certain evenings and to enjoy themselves together'. The hutty was a grey prefabricated construction about 10 m x 5 m. We met there weekly to play snooker and darts and, once a month, to get our hair cut by the itinerant barber. However, the villagers were not given the land on which it stood. Eventually, shortly after the millennium, this land acquired a new owner. The hutty had not been used for years by then and the Clatt Hall Committee had acquired it by default. The committee was given a deadline for the hutty to be removed or they would be charged for its removal.

Things dragged on until the very day of the deadline when my friend was prevailed upon to get his tractor and biggest trailer to take away the hutty. This was done in a summer evening, a few hours before the midnight deadline, amid much celebration. As the hutty was lifted off its base of packed earth, fresh soil was exposed and bugs and gases were released. The day after the hutty was removed, my friend got into conversation with a guy who had not been in Clatt long and knew nothing of the history of the area. He lived in the house opposite the hutty. Just after its removal, as he leaned over the fence watching the proceedings, he had felt a strange presence, a coldness coming from the soil of the base. It turned out that there had been a sudden death shortly before the hutty was built and the body had been found nearby.

Back in Clatt that night I shifted my feet across the soil, unaware that as I lifted the compactive force of my foot, the soil sprang back up just a little causing a tiny puff of carbon dioxide to be released. I wondered how many other people had lifted their feet from that soil. Where were the owners of those feet now? Nowadays I might also ask where was the carbon. I was brought out of my daydream and looked across at the Old Man who was looking upwards. I too lifted my eyes to the heavens and saw the lovely blue star Vega glowing right overhead and I shivered in the coolness.

The Old Man said: 'We'll go in and see how your mother's getting on.'

I looked at him and, for once deciding not to be cheeky, answered: 'OK, Dad'

Figure 3. Ploughing Insch soil at Clatt.

Figure 4. Extracting a slice of soil.

Structure quality	Size and appearance of aggregates	Visible porosity and Roots	Appearance after break-up: various soils	Appearance after break-up: same soil different tillage	Distinguishing feature	Appearance and description of natural or reduced fragment of ~ 1.5 cm diameter
Sq1 Friable — Aggregates readily crumble with fingers	Mostly < 6 mm after crumbling	Highly porous Roots throughout the soil			Fine aggregates	The action of breaking the block is enough to reveal them. Large aggregates are composed of smaller ones, held by roots.
Sq2 Intact — Aggregates easy to break with one hand	A mixture of porous, rounded aggregates from 2mm - 7 cm. No clods present	Most aggregates are porous Roots throughout the soil			High aggregate porosity	Aggregates when obtained are rounded, very fragile, crumble very easily and are highly porous.
Sq3 Firm — Most aggregates break with one hand	A mixture of porous aggregates from 2mm -10 cm; less than 30% are <1 cm. Some angular, non-porous aggregates (clods) may be present	Macropores and cracks present. Porosity and roots both within aggregates.			Low aggregate porosity	Aggregate fragments are fairly easy to obtain. They have few visible pores and are rounded. Roots usually grow through the aggregates.
Sq4 Compact — Requires considerable effort to break aggregates with one hand	Mostly large > 10 cm and sub-angular non-porous; horizontal/platy also possible; less than 30% are <7 cm	Few macropores and cracks. All roots are clustered in macropores and around aggregates			Distinct macropores	Aggregate fragments are easy to obtain when soil is wet, in cube shapes which are very sharp-edged and show cracks internally.
Sq5 Very compact — Difficult to break up	Mostly large > 10 cm, very few < 7 cm, angular and non-porous	Very low porosity. Macropores may be present. May contain anaerobic zones. Few roots, if any, and restricted to cracks			Grey-blue colour	Aggregate fragments are easy to obtain when soil is wet, although considerable force may be needed. No pores or cracks are visible usually.

Figure 5. Different types of soil structure forming the key to the spade test. Each row represents a different type of soil structure. The two rows with a green background show good structure, the row with amber background shows structure that may need improvement to realise the full potential of the soil and the two rows with red background show compacted and/or waterlogged soil which needs to be improved for sustainable agricultural use.

Figure 6. In the trenches. Revealing the secrets of the soil in profiles in the Somme valley, Péronne, France. Photo by INRA, Estrées-Mons.

Figure 7. Spiritual resources. Clatt Auld Kirk and kirkyard.

Figure 8. A potato crop pumping water from the soil. © Agroscope (Gabriela Brändle, Urs Zihlmann), LANAT (Andreas Chervet). (Chapter 3)

Figure 9. Eroded soil in Madagascar. Photo by M. Brouwers. (Chapter 3)

Figure 10. Soil degraded by waterlogging (left) and by compaction (right).

Figure 12. The deep root system below carrots. © Agroscope (Gabriela Brändle, Urs Zihlmann), LANAT (Andreas Chervet)

Figure 13. The track leading to Roadside Croft, filled with snow.

Figure 14. World hemispheres connected through soil. The author and Rachel Guimarães, Brazil observe Scottish topsoil using the spade test. Photo by Everton Blainski.

Figure 15. Different land uses: heather moor, forestry, intensive grassland and cereals together around Tap o' Noth, Rhynie.

Figure 16. Greening power. A maize seedling emerges from under the protection of the residues of the previous crop in Brazil. Photo by Neyde Fabiola.

Figure 17. Insch series soil that was cloddy and grey in the arable field (left hand side) but brown and with good crumb structure (right hand side) in the grassy area where the vegetation was natural and the soil had not been tilled for at least 10 years.

Figure 18. The giant's stone on Tap O'Noth, providing shelter for the author's daughter. The cluster of houses is the village of Rhynie.

Chapter 5

UNDER THE SURFACE

All human beings are out of their minds
Albert Ellis[1], nominee for the Nobel Peace Prize, at the age of 90

At Roadside, the Old Man was daunting. We were all wary of him because he could flare up at the slightest provocation. We couldn't tell what he was going to do from the way he looked at us. He could shout at you to get out of the way or beckon you to come close so that he could describe something of interest. I think that in both soils and human beings, a lot of activity goes on under the surface which sometimes leads to unexpected consequences. This activity happens at different levels. In soils these levels are usually defined as three layers. The top layer is about 20–30 cm deep and is usually dark because it contains humus. Humus has been described as the 'stuff of the soil'. The layer below this is the subsoil where there is less humus present so it is light-coloured, typically buff, reddish or brown. Although the subsoil is made up of either compressed sand grains or compacted clay or silt, it can usually be broken down by hand. Below the subsoil is the third layer of rock or sand called the parent material. In much of Scotland most of the parent material is a mixture of stones and sand or clay created by glaciers in the last Ice Age. Soils built up from the same parent material are grouped into families, called associations in the UK. Insch series is one of four in an association, and, like other series, is named after the village or area where most of it is found.

Just as soil has three basic layers, it seems to me that our selves can be seen as containing three layers (Figure 11). The top layer, at the surface, is the visible appearance and character of the person. This personality, the one we like to project, is strongly determined by many absorbed external stimuli, in the same way that the nature of the topsoil is affected by organic matter. The bottom or third layer is the bedrock or ground of being and embodies traits inherited from our parents and ancestors. It is also where our inner life connects with a greater reality, the universal Creative Power, Spirit, or God. Arne Naess, cited by Sarah Conn[2], calls it the ecological self – where we identify with all beings and the biosphere. This Ground of Being is the equivalent of the parent material of the soil. The layer linking the top and bottom layers, equivalent to the subsoil, is our personal unconscious: the store of our hidden potential and our hang-ups. Here are found the emotional inner images and ideas that have an unconscious influence on our actions. Nikola Patzel[3] calls this our 'inner soil'; it contains such images as the 'Mother Earth' of the Maori. These three layers of our being are similar to the three types of self identified by Alastair McIntosh[4] as the conscious self (top layer), the shadow self (subsoil) and the deep Self (parent material). The shadow self is so-called because it is what we cast on others. He states that 'for full awareness we need to bring together the conscious self, through the shadow self, into the Godspace of the deep Self'.

Plants live in two worlds, one above the soil surface and plainly visible and the other below ground and invisible. About half of a typical plant lives in each world. Below ground the root system spreads in a pattern often similar to that of the stems, stalks, leaves and branches above (Figure 12). Different crops have roots which not only reach varying depths but also have different patterns of growth. Winter wheat roots can be two metres deep, which is greater than the typical one-metre height of fully grown wheat. The roots of cauliflower can also extend to more than one metre and they are more branched, tree-like and spread out than wheat or oat roots. Tree roots go deeper, typically to seven metres.

I believe that, like plants, our existence and growth take place in two worlds, one visible, one invisible. Our physical growth is visible, particularly our sideways growth in later years! Our invisible

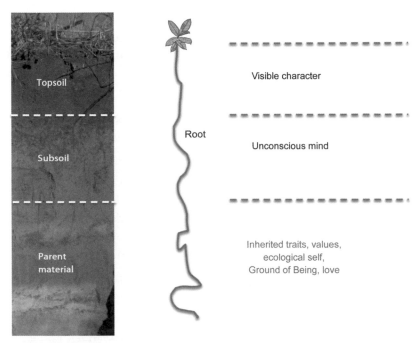

Figure 11. The basic soil metaphor. The layers of a typical soil profile (left) resemble the different depths of expression of the human self (right).

growth below the surface is shown in the development of character and spirit. We speak of having roots, which we take as being where we come from and the character we inherit from our parents and grandparents. Place is clearly important for establishing roots. Indeed, when we buy a family home we talk about 'putting down our roots'. As in soil, our roots help to stabilise and ground us within an increasingly restless society. Ideally our roots grow through our three levels of self, from the conscious self down to the deep Self. In soil the roots that grow deep are few in number but are important for supplying the plant with water when the topsoil dries out in the summer. If we establish roots in our deep Self, they will nourish our spiritual life and bring us stability and greater awareness of ourselves and our surroundings.

Many comments on the human condition use soil-related terms

like 'embedded', 'layered', 'strata', 'ingrained', 'shallow', 'deep' and 'coalescence'. These words suggest complexity and inflexibility and indicate aspects of our personality that may be hidden. The things that we say are often not a reflection of what is actually within us. Observation of minor details of behaviour may give a better insight into a person's real nature than his or her conscious actions and expressions. This is shown in the social games that we all play, the little rituals and behaviour patterns between individuals which often reveal hidden feelings or emotions. For example, the Old Man would play 'If it weren't for you I could …'. He would blame his parents for not having given him a chance and also complain at us for holding him back, preventing him from making more money and achieving success in his plans.

Nevertheless, the Old Man was clearly in control at Roadside. As we settled down for the evening, he sat in the only easy chair in the main living room and the rest of the family fitted in round about him. I used to sit on a tiny stool close to the fire, wedged between the bed and the Old Man's chair. That was cosy, but you were always aware of his domination; it was his word or none. Even though he didn't have much going for him, he had the power – his will was done. However, his lifestyle was inflexible. He was a creature of habit, isolated by his inability to get on with others. My mum was the one who went out in the evenings to meetings of the Women's Guild and the Scottish Women's Rural Institute. She made the friends and she held us together. She developed the networks in a spirit of conservation, cooperation and partnership.

It was curious how the Old Man's power and control thrived only in its own environment. If you took him away from his power base at Roadside, then he behaved more reasonably, though if we took advantage of this, we often paid for it on our return home. The system that we had at home can be seen as part of larger systems driven by oppression, well-defined power structures and male domination that are often maintained by the subtle use of violence or threats. These include the church. At Clatt, the laird was clearly in control and appeared to use Clatt Kirk to reinforce his power base. This was suggested by the way that he walked into the church. He was a big man, and the swing of his kilt and the fling of his wife's fox fur around her neck told of landed power. He sat in the front pew right in front of the little pulpit, the best seat in the house. It must have been nerve-wracking for the minister to be faced with this Sunday by Sunday.

Many years earlier, in 1571, a bloody struggle for domination had been fought out on the pleasant fields of the parish of Clatt. This was at the time of the civil war between the supporters of Queen Mary and her son, King James VI. The Catholic Gordons, wishing to gain the road to Edinburgh – the one which passed Road-side – were opposed by the Protestant Forbes and a battle occurred near the farm of Tillyangus with the Gordons proving victorious. It appears to be an early episode of sectarian violence.

If we look at the soil we see a system without power, but with strength. Strength is provided both by the organic matter glueing together each aggregate and also by the support of the surrounding aggregates. As Neale Walsch[5] states about the human condition, it's not about separation but about unity. It is in unity that inner strength is found and in separation that it fragments, leaving individuals weak and struggling for power.

Harry Palmer[6] explains that the core beliefs which drive our behaviour and our attitudes to others are also hidden within us and may not even be known to us. One of these hidden myths relates to power and belief in the ability of violence to overcome violence[7]. In common with most other boys, we played out conflicts at Clatt, usually in the form of cowboys and Indians around the farm buildings. There were plenty of places for the Indians to hide, who, at the appropriate sign, would be hunted out by the cowboys. The cowboys had pretend guns for which the bows and arrows of the Indians were no match. As the Indians raced for cover and heard 'Bang, you're dead', they were soon all lying on the earth. The cowboys won by killing all of the Indians every time so that order was restored. We just accepted that this was the case.

I later found out that we were acting out a common myth that violence can be overcome by violence, provided that the technology is more powerful and the perpetrators more 'civilised'. From early in our lives we are exposed to cartoons and TV shows which follow the same basic plot. This goes: (1) a bunch of baddies create general havoc, (2) a hero comes along, fights these baddies and either kills them or suppresses them for a while, and (3) peace and harmony are restored – resolution through violence. Remember Tom and Gerry, Power Rangers, Batman? This ingrained thinking carries over into our willingness to sort out problems by the use of violence and war – or, in the case of nuclear weapons, by the threat of violence. As we have observed so many times recently, this power-centred structure of society is ultimately self-destructive.

Another more obvious myth that we live by is that the getting, having and using of material things produces happiness, satisfaction and hope. In Clatt, this was well displayed in the realm of personal transport, particularly among the males. Living in an isolated place, the villagers depended heavily on private transport. As kids they started with a bike, at 16 they progressed to a motorbike, then at 18 to a small second-hand car, through to a larger second-hand car, then a small new car, to a large car, and then to two cars and so on. There was a clear progression of increasing size, novelty and number. Each item was usually lovingly washed and cared for.

This is by no means unique to Clatt; it is a phenomenon found just about everywhere where there is disposable income. It of course applies to other possessions as well, such as clothes, electronic goods and collectors' items. Tim Jackson[8] considers that the accumulation of goods, often as a result of addictive behaviour, can be seen as part of our shadow self or subsoil which is ultimately empty; it cannot be filled, no matter how many consumer products or celebrities we feed it.

Some of you will remember the winter of 1963 when Scotland was covered in snow for many months. This was our first winter at Roadside and the heavy snowfalls and biting, persistent winds produced huge snowdrifts and a dazzling landscape which bore no resemblance to the actual landscape beneath the snow. In this way, the normal landscape became the landscape below. Our access track to the cottage was lower than the level of the surrounding fields (Figure 13). It was filled in completely and the fences, the stones and the gate were hidden by a shiny, smooth, pure white surface. This landscape revealed where you had walked and also told of the passage of time in a way which I only understood when I found one of the Old Man's poems after he died:

I've walked down the snowy path
Through winters ten or more
I've noticed where small feet have trod
O'er shining snow so pure

But as the winters later came
As father I could see
The smaller feet marks, they were gone
That years ago were wee

After one or two days of frost, the snow was hard enough to walk on without sinking in, so that we could walk between fields without being obstructed by fences and could climb temporary ridges provided by snow drifts. Great for a ten-year-old! However, as spring came, it all melted away to show the true landscape, which looked dark and grey and had potholes at the surface. In a similar way, I believe that we sometimes deliberately hide things below a shallow veneer which may look smooth, but is a thin and temporary facade for what is below. A guise of respectability may hide problems such as addiction to alcohol or drugs, or abuse.

Nevertheless, like the revitalisation of the soil that comes with the spring, good things can also emerge from under a rough surface. There are wonderful hidden abilities and talents within most of us, some of which we often seem disinclined to use, like fertile soil never used for growing crops or trees. There are some excellent leaders, musicians and carers out there who just don't do what they are good at.

Talent is also often hidden because of poverty or deprivation. If people in poor areas were to be trained and empowered, their hidden abilities would be released. This is especially important for poor people in developed countries, who represent 'undeveloped strata'. Community spirit may also be hidden below the surface. When working as an aid organiser in a deprived area of Chicago, Barack Obama[9] found that no matter how tough the individuals were, there was a community there if you dug deep enough. He spoke of 'a luminous world' beneath the surface of the community. To access it, all he had to do was ask.

When a farmer applies organic matter to the soil, worms and moles multiply and thrive so that soil of good quality comes to the surface in the form of worm casts or molehills. Earthworms are remarkable in the way that they take in organic debris and soil, and mix them together with their digestive juices in a grinding action that produces a paste which is excreted as a cast. Casts consist of small, stable aggregates rich in available nutrients and humus. Most earthworm burrows in soil are invisible because they are filled with these casts.

I remember taking soil samples from a field experiment near Roslin which had been left uncultivated for a long time, setting them in translucent resin and cutting very thin slices called sections (about 8 cm deep by 5 cm wide). I showed them to an expert whose first comment, after looking at them through a microscope, was that the entire soil was made up of worm crap (casts). The humble earthworm is a remarkable shaper, creator and maintainer of the landscape below. Earthworms do so much for us, sight unseen, that it is easy to forget their importance. Harry Edmund Martinson[10] expresses this well:

Who really respects the earthworm,
the farmworker far under the grass in the soil.
He keeps the earth always changing.
He works entirely full of soil,
speechless with soil, and blind.

He is the underneath farmer, the underground one,
where the fields are getting on their harvest clothes.
Who really respects him,
this deep and calm earth-worker,
this deathless, grey, tiny farmer in the planet's soil.

I've never seen a molehill being pushed out of the ground. Although a nuisance to the farmer, they are a sign of fertile soil. In a similar way, if by training and encouragement we foster growth or the release of hidden feelings in others, fruits of the spirit appear from below, often without warning, as little deeds of love. Time spent listening to someone and allowing them to work things through may result in a card in the letterbox or a job done without asking.

Human spiritual activity is hard to detect and many refuse to accept that it even exists. Such action can be hidden within communities but is nevertheless very powerful. I now live in the village of Roslin which is famous for its chapel, as publicised by Dan Brown in his novel *The Da Vinci Code*. Many think of Roslin as a spiritual place because of the chapel. The hidden spiritual activity that I'm aware of in Roslin does not relate to the possibility of discovering the Holy Grail in Rosslyn Chapel or indeed the spirit within the chapel itself. No, it lies in a street which the tourists ignore as they stream down to the chapel. This, as described by Ron Ferguson[11], contains the location of a dilapidated former miners' welfare institute and a few huts behind it, including a little chapel. At the time of writing, a handful of people still come to visit and worship in this chapel at the Community of the Transfiguration. These people seek guidance, help and healing. It is a tiny community which gives hope and inspiration, and its work, though hidden in Roslin, has grown and multiplied in other communities, mainly as part of the Franciscan family.

The spiritual events stemming from Roslin and from within other communities, organisations and individuals are often almost invisible but have a profound influence on many lives. This kind of action of the Spirit is nurtured by developing our connections to each other and to nature and agriculture, so that we become fully aware of our integral part in the network of life.

Now

Know the now
Hear and here
Live the moment
Keep the gift
Of the present
All the time

Chapter 6

GETTING CONNECTED

This is a membership one of another –
community of soil, soul and society
Alastair McIntosh[1]

The Clatt farmers, in common with most, feel a strong connection
to the land. It draws them towards the farming life, and perhaps
explains why farms are handed down from generation to generation.
Working on the land provides food, water and recreation and creates
an emotional attachment, especially if it has been passed down from
your parents and grandparents. The land is part of you, supplying a
living link back to your ancestors. People find it hard to survive
when separated from their land, nor can the land survive in a productive state without those who are naturally part of it. Workers of
the land are aware of coming from the soil, of being sustained by it,
of the need to care for it and of eventually returning to it.

Understanding the connection between soil and food is vital.
Many children are unaware of the source of their food. We increasingly live in cities where we are remote from contact with the land
or cultivation of crops or rearing of animals. Moreover, city and
town dwellers tend to be cremated rather than buried, which breaks
the cycle of returning to the earth. All of these factors serve to lessen
our link with the land. Fritz Schumacher[2] saw that one of the main
tasks of agriculture is to keep people in touch with living nature, of
which they are a vulnerable part.

I think that love of the land is something that grows within us. I
love to handle the soil and find a peace of mind in doing so, yet I
have known other soil scientists who do not feel this affinity. They
tend to treat the soil as a mixture of separate components – chemical, biological and physical – which can be measured individually.

While the methods of these scientists are effective, I prefer the holistic approach, which treats the soil as a living organism. In this way the love of the soil can be understood, a love born of contact, particularly during crop growth.

Although the soil is most exposed when it is freshly ploughed, it is most vulnerable after it is cultivated to produce a seedbed. The soil aggregates are small and the surface is usually level and fairly firm. The seed and fertiliser are then sown a few centimetres below the soil surface. Everything connects together in hope when seeds are dropped into the soil and everything is at risk at this stage. The seed may be eaten by birds, the soil may be eroded and the fertiliser may be washed away by rain or blown away by the wind. At this time, the connection between farmer and soil is at its strongest and the call of nature is most keenly felt. The vulnerability of the soil just after sowing reminds me of our own hopeful yet vulnerable state during transition periods in our lives – times of birth, marriage and death for example.

Normally, within a few days a yellow bloom comes over the field, which rapidly turns green as the crop emerges. This is the 'greening power' of divine creative energy, first recognised by Hildegard of Bingen[3]. The soil soon becomes less vulnerable, held together as it is by the fast-growing roots of the crop. I never lose the sense of wonder at the miracle of creation when the crop appears in the field or in the garden. This sense of a 'greening power' and other emotional images of the power or spirit of the soil such as 'Mother Earth' or 'vital force' were described by Nikola Patzel[4] as 'inner soil'. He believes that such archetypes live within the unconscious of soil scientists, driving and guiding their actions. If this is the case, developing the inner life is clearly important for innovation in soil science. He also believes that our loss of respect, wonder and reverence for the soil has contributed to the accelerated and ongoing soil deterioration and destruction in many parts of the world.

Our connection to the soil is established not only by sight but also by the other senses, which need to be reawakened to renew our bond with the earth. Many people learn about the soil by walking on it, especially when it's soaking wet and good shoes are being ruined. I reckon that by walking across a harvested field I can judge what type of tillage is needed to establish the next crop. Hard, smooth areas or areas with wheel ruts will need tillage, whereas softer areas that sink a little beneath the feet and crumble around the stubble of the previous crop may need little or no tillage. Barbara Taylor[5] suggests that standing or walking on the earth with bare feet is a spiritual practice, because you feel the world as it really is and you can experience it truly as holy ground.

In the spade test of soil structure, a lot can be learned from touching the soil. The ease with which it breaks up is critical to assessing the quality of the structure. Handling aggregates is important. Those that feel rounded and light in weight are usually good, whereas sharper, flatter, heavier aggregates are not so good. This kind of connection to the soil is like listening to music: there are no barriers of language or custom (Figure 14). Perhaps the ultimate and most primal way to make contact with the soil is to roll in the mud and rub it over yourself, so that you are in effect absorbing the land through your skin.

Rebecca Lines-Kelly[6] believes that touching the soil literally earths us; that we have a subconscious link to the soil that goes deep into the core of our being and connects us with the spirit of what it means to be alive. She also quotes Wendell Berry, who considers that anyone contemplating the life of the soil will come to see it as analogous to the life of the spirit. Many have analysed and worked with soil for years, yet it is such a complex organism that we still have a very poor knowledge of it. Stephen Lewandowski[7] thinks that soils are wild in that they are unknown territory, self-regulating and beyond our control – just like Spirit.

If the focus is fully on the essence of what the soil has to reveal through us, it is also possible to engage the Spirit when down a soil pit. Here, words which sound good in a warm lecture theatre accompanied by nice, carefully selected PowerPoint images seem less convincing as the rainwater runs round your mouth or off your soil-filled hand and onto boots heavy with subsoil.

After handling the soil, I try to avoid saying that my hands are dirty or soiled. In the Maori language, there are no words for this. Nevertheless, the hands are no longer clean: particles stick to the skin and the organic matter leaves dark markings. I feel uncomfortable, a bit contaminated, so I wash them or rub them in wet grass. Perhaps it is not really necessary. Rebecca Lines-Kelly believes that if we rid ourselves of the mental connection between soil and dirt or excrement we may improve our connection with the soil.

Children can be more susceptible to illnesses in childhood if they are not allowed to play in the mud or are forever being made to wash their hands. As a nation we certainly like to keep ourselves clean. Some who are over sixty will probably remember the time when a bath was taken only once or twice a week. Now many people shower every morning. I wonder why we do this, especially as our hospitals and public places are dirtier than in those days when we were 'less clean' personally. Alastair McIntosh considers that 'the great washed' are the ones we need to reach out to with spiritual help. If we washed a bit less it would save precious energy and water too.

We also connect to the soil through the sense of smell. I sometimes teach the spade test in lecture rooms or village halls where I will have up to twenty slices of soil spread out on tables ready for handling. If the room is kept closed, the soil as it breathes produces a strong odor, making all aware of its life. When doing the test, I encourage people to smell the soil. They soon learn the difference between the mellow, clean, sweet smell of a soil rich in organic matter, the fainter, earthy smell of most agricultural soils and the yucky, sulphurous smell of soil that has been waterlogged and is in poor condition.

Another connection to the soil which is perhaps less obvious is through hearing. I've already mentioned the sound of soil being cultivated. Other direct sounds occur after heavy rainfall, especially in cereal stubble. As the water drains out of the very large pores in the soil and they fill with air, bubbles make their way upwards through the pore system. As they burst at the soil surface, they often produce a series of plops. There are also other, less obvious sounds – the speech of the land as described by Sheena Blackhall[8] in part of her poem 'The Spik o' the Lan':

> *The chap o' the preacher's wird,*
> *Be it wise as Solomon,*
> *It fooners on iron yird*
> *Brakks, upon barren grun.*
>
> *Bit the lowe o' a beast new born,*
> *The grieve at his wirk,*
> *The blyter o' brierin corn,*
> *The bicker o' birk*
> *The haly hush o' the hill:*
> *Things kent, an at haun*
> *I'd harken to that wi' a will.*
> *The Spik o' the lan!*

Peasants in ancient Slavic tribes used to listen to what 'Moist Mother Earth' was telling them by digging a hole in the ground with their fingers and placing an ear to it. James Swan[9] considers that listening to this language was part of their worship of the earth. He also believes that there is a voice deep within us that understands nature – and our own nature – and that we should listen to its promptings. Thom Hartmann[10] describes how the Shoshone native North Americans located roots and food in the ground by listening to, communicating with and looking at the visible and hidden landscape.

The last, and least likely, connection with the soil is through taste. You are what you eat and what you eat is of the earth. This was brought home to me when I visited Shaanxi province in China in 1990. I went to the remote North-Western Agricultural University at Yangling, near Xi'an. This was just after the student riots in Tiananmen Square in Beijing. In that remote area, the lifestyle had been little influenced by the West and the food was still very much locally sourced. The soil type is yellow-brown loess with a parent material of yellowish, soft, silty sediment blown over from the desert of Central Asia, making it one of the deepest soils in the world. The particles of soil tend to get everywhere – on your hands, in your clothes – and, to my surprise, gave the food, particularly the fruit and vegetables, a distinctive earthy taste. I was glad to leave this taste behind when I returned home. During later travels, I found that the local food cooked in Maori hangi in New Zealand was more tasty. A hangi is made by digging a hole in the ground, lining it with red-hot stone and putting on a layer of vegetation. The food, wrapped in cloth, is placed on the top, covered with more vegetation. The pit is then filled with soil and the meat and vegetables are left to steam.

These connections to the soil – or any other being or thing – are, I believe, enabled by 'Spirit' and increase our awareness of the environment, motivating us to care. Spirit is often seen as a religious property, but many have experienced it who do not associate it with religious belief. My early realisation of being as one with stone in the garden and with the soil during tillage resulted from my being fully aware of the present moment. I was living in the now and experiencing 'presence'. This is the basis of spiritual awareness and is the goal of meditation. It demonstrates the fundamental difference between spirituality and religion. Spirituality is discovery by oneself whereas religion is, often, learning from the experience of others.

The memory of the 'presence' never went away. I had established a connection with the 'Spirit' in the landscape. The belief that Spirit is within everything, not just living creatures, and yet is transcendent, is called panentheism. Places and communities also have this Spirit. The spirit of a place arises from features such as its location, size, topography, the presence of water, and its remoteness. Moreover, if we accept Lovelock's description of the Earth as a living super-organism, Gaia, it too has its own Spirit. We are part of this Spirit and are held within the biological system of the Earth rather than merely living on its surface. Leonardo Boff[11] believes that this means that everything is inextricably interconnected to every other particle of matter in the cosmos in a common consciousness which extends beyond the deep Self, or parent material, to include the relationships between persons, continents and cultures. Thus, the entire cosmos can be seen as a super-organism.

The astrophysicist Bernard Haisch[12] extends this idea further to describe a being which is limited to neither space nor time. He considers this unlimited conscious being, or Spirit, to be God. Earlier cultures, too, saw the earth as the embodiment of a great Spirit, the creative power of the Universe, present in all things – rivers, trees, mountains, springs and caves. Alastair McIntosh[13] describes this as the 'essential Ground of Being'. It brings home to us that we are beings of three facets: body, mind and spirit.

Spirituality is defined by Brian McLaren[14] as having four basic characteristics. First is acceptance that life has a sacred dimension that cannot be reduced to formulas, rules and numbers. Second, spiritual people have an inner sensitivity to aliveness, meaning and sacredness in and throughout the universe – a universal sense of the integrity of everything and everyone. Third, this feeling of aliveness needs to be maintained by practices such as meditation or worship. Fourth, organised religion does not have all the answers. This all boils down to 'that which gives life' (Alastair McIntosh) or 'seeking vital connection' or, in a single word, love (Brian McLaren).

People and place define spirituality. Spiritual aspects include ideas of truth, wisdom and love. When the spiritual connection is good, then we learn to recognise our 'gut' instinct or intuition and when to go with it.

The idea of Spirit for us in the West has mainly anthropomorphic or human associations, whereas indigenous cultures, according to D. Abram[15], consider 'spirits' primarily as modes of intelligence or awareness that do not possess human form but are nevertheless part of Spirit. Ancient indigenous peoples have developed an awareness and respect for the Spirit which they express in their everyday life and in their approach to living with their environment. For example, in the farms around Clatt, an old tradition was to leave an area of land called the 'Guidman's Ground' uncultivated or ungrazed out of respect for the forces of nature. It was widely believed that cultivation of this ground would bring misfortune, notably in the form of cattle disease. The Church actively discouraged this practice as they thought that the land was dedicated to the Devil. However, the belief was very strong and the Church found it hard to impose cultivation of the land, despite heavy fines.

Engaging the Spirit for reconnection often involves the prefix re: re-generation, re-alisation, re-vision, re-clamation, even re-ligion. Restoring our connection to the soil is similar to restoring other vital connections which we are in danger of losing. Technology has shrunk the earth and provides great opportunities for communication. Yet somehow, in spite of all this, we seem to be losing our connection or connectedness with each other and are becoming more isolated. More and more of us live on our own. When we travel, we are remote from both road and landscape in our increasingly large and comfortable cars. Isolation is becoming more common in spite of improved road networks. For example, in Scotland's Western Isles the almost universal use of cars for transport discourages neighbourly visits because of the risk of breaking drink-driving laws. The Clatt people used to walk, cycle or drive to the local shops for food where they would meet and pass the time of day with each other, but now the local shops have gone and most people drive the 12 miles to the large supermarkets in Huntly. Even when we walk we isolate ourselves by listening to music through in-ear headphones.

We communicate more and more by email, mobile phones and text messaging, and through social networking sites. At least we can hear each other when we converse using mobile phones. Social networking sites provide a temporary, unsubstantial web of friendships. Alastair McIntosh believes that we also tend to share community via TV rather than in real life – witness the popularity of the UK 'soaps' Coronation Street and EastEnders. But would you actually want to live in such strife-ridden communities? TV is strongly isolating and provides information that is remote from our own existence. Another powerful, but ultimately destructive, means of communication is the language of goods. We identify with those who wear the same type or brand of clothes or who drive the same model of car as we do. Tim Jackson[16] believes that the goods become part of our 'inner soil' or shadow self. However, just as good topsoil cannot develop on waste or sterile material, this shadow self is ultimately empty.

We all depend on a web of relationships with one other to survive and it is only possible to know ourselves against the background of those relationships. Authentic relationships are developed by good communication, which is best established face-to-face. Even there, communication takes place at varying levels, from a superficial chat about the weather to deep exchanges when we start to talk about things that really matter to us.

When, for this book, I was talking to my farmer friends in Clatt about their relationship with the land and with the soil, I was surprised at the warmth and depth of the communication. Most were older than I was, and such connection with our elders, honouring and respecting them and listening to their wisdom, helps to establish a continuity of life. Reconnecting with the past, and with our ancestors, increases our sense of personal and collective identity and responsibility which we, in turn, pass on to others, especially our children. This sharing of words reminds me of how earthworms create continuous pores from the top layer through to the lower subsoil, connecting the soil of the present to the old, undeveloped subsoil and improving quality by helping air to enter and water to drain away.

People are no longer bound together by the rites and customs that give a sense of belonging in traditional culture. In an established community, when someone talks about a person unknown to you, they will first lay out the family background ('He came from Clatt and married so-and-so, his second wife, so I believe' etc.). Although this may seem like gossip, it serves to make clear the connections within the community and is time well spent. This realisation of 'family' connection may help to develop awareness of the obligations of future generations for environmental sustainability. Daniel Hillel[17] considers that the ultimate purpose of environmental activity is to make sure that each generation transmits to its successors a world that has the same (or a better) range of natural wealth and riches of human possibilities as it received from its predecessors.

Re-establishing connection is also possible through sharing in small groups, as in early tribal communication. I was involved in the development of house groups in the Church of Scotland back in the early 1980s. Such groups allow people to speak to each other about their hopes, problems, fears and joys without the need of input from a professional such as a social worker or church minister. Alastair McIntosh[18] thinks that such sharing can lead to a deep communication, enabling our roots to extend below the grass roots of popular culture down to the deep Self, the ancient stock from which we grow. Of course, we can also find connection on our own by spending time in quiet appreciation and meditation. There are many ways of personally engaging the Spirit, such as through simple meditation to increase self-awareness and give a sense of peace and purpose, and through prayer, which develops our sense of compassion, gratitude and hope. There are plenty of other methods of allowing the soul to express itself – writing poetry, making music, talking till dawn, singing in the rain, etc …

Our connection with the environment is broken by our constant exploitation of it. We have lost a sense of respect for that which is sacred, holy and mysterious in our surroundings. Leonardo Boff[19] describes the sacred as that quality that fascinates us, speaks to us of the depths of our being and gives us the experience of respect, fear and reverence. Thom Hartmann[20] describes every bit of Creation as being alive, vital and sacred – having spirit or soul – and so touching us with love. The teachings of Buddha regard Earth itself and all of its life forms, right down to the very lowest, as sacred. Stephen Lewandowski[7] believes that restoring the sacred to the land involves applying the principles of sustainable agriculture, some of which are ancient.

We hear regularly at funerals the words 'ashes to ashes' and 'dust to dust', reminding us that we are from the earth and will return to it. Leonardo Boff[19] believes that when we become conscious of ourselves as earth, we start to feel at one with Creation and recognise that part of the mountain, the sea, the air, the tree, the animal – of the other and of God – is in us. Those who espouse the philosophy of deep ecology regard this consciousness of being connected to the whole – which I call Wholeness – as the natural human state.

The Bible has had a great influence on how we use or abuse the resources of the earth. Daniel Hillel[17] reminds us that there are two accounts of creation given in Genesis. The first, the one we are most familiar with, is in Genesis 1:27 (NIV): 'Be fruitful and increase in number, fill the earth and subdue it.' This image is still conveyed by many followers of religion. For example, Christianity usually teaches that life is a straight-line progression. It is important to go beyond this to thinking in cycles. Further on, in Genesis 2:15 (NIV), after God formed humans out of the soil of the earth, there is a second account: 'The Lord God took the man and put him in the Garden of Eden to work it and take care of it.' Human beings were charged with the responsibility of nurturing and protecting God's

Creation. The often-ignored second account is surely our proper role in relation to nature. Ron Ferguson[21] sums it up well: 'The whole earth is sacramental ... Reverence for the earth, God's sacrament, is not only right and fitting, it is essential for the survival of the planet. The image of man as the dominant exploiter of the earth must be replaced by that of man as steward of God's creation, holding all things in trust.' We are an integral part of the environment, not masters of it.

Connection with the environment requires some spiritual communication. The common spiritual source for many in Scotland is the Church. However, the traditional messages of the Church are often difficult to follow and act upon. Two of the main reasons for the lack of interest in the Church at Clatt, and elsewhere in the Western world, is the perception that it is no longer spiritual and that its powers are misdirected. People turn to other means of developing their spiritual life such as meditation, yoga, witchcraft and card-reading. It is important to bear in mind the distinction between spirituality and religiosity. Bernard Haisch[12] considers that religions differ because they are bound up with culture, whereas spiritual truth is universal. Brian McLaren[22] suggests that religion 'can become a benign and passive chaplaincy to a failing and dysfunctional culture ... it can forgo being a force of liberation and transformation and instead become a source of domestication, resignation, pacification and distraction.'

Religion seems to have lost the drive of the revolutionary. For example, Jesus' main concern was for the poor, the vulnerable, the marginalised, the children, the forgotten and the oppressed. St Francis of Assisi shared these concerns and saw all other living things as our brothers and sisters in the history of the earth, inextricably connected with us and our life source. Alastair McIntosh[23] considers that this connection is at such a deep, psychic level that ignoring the suffering of the world violates our very nature. Crucial to this is loss of the connection between rich and poor, resulting in injustice. This connection can be restored by concentrating on other people, other creatures and the environment. These form one Wholeness through the spirit. Every action we take influences the integrity of that Wholeness. Awareness of this Wholeness brings peace. When I stand in the churchyard at Clatt (Figure 7), surrounded by those who went before and with the solid church at my back, I become aware of how the crops, the fields, the woods, the hills and the homes all fit together. It gives me a good feeling inside and at that moment I don't worry too much about what will happen to me.

If agriculture and other uses of land and resources are to be environmentally sensitive, they need to include spiritual and metaphysical aspects. Moreover, a spiritual awareness will influence our role as consumers, whether the system in question is organic farming, land restoration, conservation agriculture or another.

Chapter 7

GOING ORGANIC

The birthright of all living things is health.
This law is true for soil, plant, animal and man:
the health of these four is one connected chain.
Sir Albert Howard, 1947[1]

As a student I worked on several farms around Clatt. My transport was an old motorbike I had bought from the retired gamekeeper of Knockespoch estate. It was a powerful Norton 500cc machine that had only done a few thousand miles during its 12 years. It came with an old-fashioned sidecar attached – soon removed – and, with a bit of a clean up, the bike looked as good as new. It was great to drive, and gave me a real sense of freedom as I navigated the narrow roads around Clatt. Going to work through the main street, I would wake the village up shortly after 7am. As I wound open the twistgrip throttle the 'thud, thud, thud' of the big single-cylinder engine would switch to a more urgent 'thwack, thwack, thwack', reverberating off the walls of the houses. I was on my way to work at a farm near Insch where I cleaned out pigs, turned hay, loaded bales, hoed turnips and brought in grain and silage. I enjoyed the work and the family there made me feel most welcome, even feeding me. Partly as a consequence of being in the house for meals I would sometimes end up doing non-agricultural jobs like weeding the garden or washing the car but I didn't mind as it was a great place with friendly people and a relaxed approach. The spirit there was good.

A year later I worked at another farm near Huntly. This was a bigger, more intensive enterprise consisting of several farms worked together. I did long hours, earned more money and could afford a car, but the spirit wasn't so good. It was run by a manager, so I rarely saw the farmer and there was much less sense of camaraderie. One of my jobs was to look after several hundred pigs in a huge barn where the noise at feeding time was deafening. It was like working in a concentration camp for animals. I was also part of a 'flying squad' where a team of tractors and machines would move from farm to farm establishing crops or making silage. This was hard going and I found it difficult to keep up. I didn't like it there and soon moved on to another place which was almost as big, but friendlier.

The three farms all produced similar crops, but the management of the resources and the workers was quite different. Working on land where the farmer still lived in the farmhouse resembled the traditional farming role of supporting the family, whereas working the land where no one lived was more like exploiting a resource for profit. I believe that this influenced the spirit of each place and the quality of the food that was produced. Wendell Berry[2] considers that technology allows farming to increase to a scale that is undemocratic because there are not enough people owning and working the land and there is insufficient attention to the overall quality of the farming system.

None of these farms was organic because the modern idea of organic farming was uncommon in the UK in the early 1970s, though the farm near Insch came close. I would prefer to eat produce from this farm even though analysis would probably show no differences in the level of nutrition of the food between the farms. Many people prefer organic food. It is generally regarded as more healthy and tasty than conventionally grown produce, but perhaps the circumstances of its production have an influence on people's choice as well. Taste is difficult to quantify, but the UK Food Standards Agency in 2009 and Stanford School of Medicine and University in 2012 concluded from independent reviews that there was no evidence that organic foods gave either better nutrition or health benefits. It's hard to define the spirit of the food that you eat.

Organic agriculture is based on four principles, as listed on the IFOAM website, of

- *health* – sustaining and enhancing the health of soil, plant, animal, human and planet as one and indivisible.
- *ecology* – based on living ecological systems and cycles, working with them and helping sustain them.
- *fairness* – building on relationships that ensure fairness with regard to the common environment and life opportunities.
- *care* – precautionary and responsible agricultural management to protect the health and well-being of current and future generations and the environment.

I wish that our lives were run on the same principles. The organic system is an accredited farming system which restricts chemical additions to the soil and medicines for animals and is strictly controlled by regulations defined by the International Federation of Organic Agriculture Movements (IFOAM) and, in the UK, by the Soil Association. Organic farming is commonly seen as better than conventional farming, being on a smaller scale and carried out by locally based farmers. It should improve the ecological stability and long-term environmental sustainability of farming because less fossil fuels and chemicals are used and it utilises more beneficial natural processes and renewable resources that are available on the farm.

Farming at Clatt was still almost organic when I first lived there, with fertility maintained by crop rotation and animal manure. Rotations included clover, a legume crop that captures nitrogen from the atmosphere and stores it in its roots. This was ploughed in, and as the roots decomposed they released the nitrogen, which nourished the arable crops. Animals, usually cattle and sheep, fed on the grass, clover and grain to provide manure for the arable crops. These are still the basic methods of cycling nitrogen nutrition on organic farms.

Unlike organics, mineral fertiliser containing potassium and phosphate had been applied on the farms at Clatt since the early 20th century. Nitrogen began to be used later in the 1950s in compound chemical form, and nitrogen-only fertiliser appeared in the 1970s, a period during which crop yields increased greatly. The ability to create nitrogen fertiliser from nitrogen in the atmosphere is one of our greatest inventions. Without it we wouldn't be able to sustain our population. Initially, there was resistance by some farmers to the use of nitrogen fertilisers, with claims that they would 'suck the ground'. Nitrogen-only fertiliser certainly acts quickly and stimulates leaf growth. However, I think that the way it nourishes the soil is similar to the way we use sugar. Both provide energy and growth, but not balanced nutrition.

Nitrogen fertiliser comes with other problems too. The equivalent of 170 litres of fossil fuel is required to produce the fertiliser needed for one hectare of agricultural land in intensive cereal production. Our use of it is so inefficient that as little as 5% of the nitrogen in the fertiliser actually ends up in our food, with the rest being lost to the environment. All of this is pushing us into an unsustainable position globally because of the huge environmental effects of poisoning of lakes and seas by eutrophication and increased greenhouse gas emissions. To me, the use of high levels of nitrogen fertiliser is like the use of large tractors and pesticides. They all represent our desire to dominate and control modern agriculture. Indeed, inequality, the gap between rich and poor, has been considered to have arisen and spread with the development of agriculture[3].

Nevertheless, to maintain and increase productivity, agriculture will probably need to continue to use nitrogen-only fertilisers, although a switch to compound and slow-release types of fertiliser or to a rotation of crops involving legumes would mean greater use of nitrogen by the crop with fewer losses to the environment. Since mineral fertiliser cannot be used in organic systems, organic farmers rely heavily on the fertility provided by nitrogen fixed in legume crops such as clover, alfalfa, peas and beans. Organic rotations usually contain several different types of crops and animals. All this, along with the challenge of controlling weeds, pests and diseases without chemicals, demands a high standard of management and commitment for organic farming to succeed.

Modern organic farming can be considered to have a spiritual history. Its early pioneers spoke of Mother Earth, implying a link with 'inner soil' and feminine wisdom. They also emphasised the living nature of the soil, usually the organic matter or humus, and the cycle of growth and decay. Organic farming was originally, and still is, seen as a holistic system which assumes that a farm is a single, whole, complete organism. If any part of the system changes, then the whole farm changes. This unity or holism was believed by Sir Albert Howard[4] to be important for health. Wendell Berry[2] connected this with spiritual, economic and political health. One of the early pioneers of organic farming in the UK, Lady Eve Balfour, considered that this holism means that 'we cannot escape from the ethical and spiritual values of life for they are part of wholeness'[5].

In Germany, organic farming grew out of anthroposophy, as described by Rudolf Steiner. Anthroposophy is a spiritual philosophy based on belief in the existence of an objective, understandable spiritual world accessible by direct experience through inner development. Steiner was convinced that mineral nitrogen fertilisers could not produce healthy food and developed an interest in organic farming.

One of Steiner's legacies is the curious version of organic farming called biodynamic farming in which spiritual aspects are included in the day-to-day management of the land. Biodynamic farming also takes into consideration the influence of planetary rhythms on the growth of plants and animals. Soil renewal is an important feature and is achieved through special composted preparations used to treat the soil and plants. Materials are fermented in animal body-parts such as cow horn which, it is claimed, help concentrate the life forces from the surroundings into the material within. I visited the research centre for biodynamic farming at Darmstadt in Germany in 2001. The spirit there was good and bright. I learned about the claimed benefits to growth of preparations of silica and herbs incubated in cow horns buried in the soil and then applied as sprays to the fields. An interesting spray is the Three Kings Preparation, a mixture of gold, frankincense, myrrh and glycerine, which is put on to the perimeters of a farm or village on Three Kings Day (6th January) to ensure growth later in the year. Recycling, composting and the integration of all forms of life are particularly important in biodynamic farming. The approach is holistic, with the emphasis on the connection between soil, crop, animal, producer and customer. I was impressed by the good overall grounding in life skills being given to the students at Darmstadt. They were taught how to paint as part of their course and spent quite a lot of time outside in the presence of the crop, talking to it. I'm not sure how much of a direct effect this had on growth, but time spent amongst the crop and becoming aware of its needs will make a better farmer.

Soil conditions affect our health more than we are probably aware. For example, prospective parents in the UK may not realise that choosing a house in an area where the soils are dry can increase the chances of their baby surviving. Ensuring that the youngsters have regular contact with the soil can increase the microbial diversity in their stomachs and possibly reduce their chances of developing allergies. As the parents grow older they may wish to avoid living on poor quality, acid soils which may (slightly) increase their chances of developing heart disease. Sir Albert Howard[4] predicted in 1947 that the 'public health system of the future is based on the produce of land in good heart'. His predictions seem to be coming true. The trace mineral content of fruit and vegetables is in long-term decline. Trace minerals are sodium, calcium, iron, magnesium, copper and selenium; they are essential for our health and well-being. Their decline means that our five-a-day of fruit and vegetables packs less punch than it used to. We rely on getting minerals from our food; we can't make them ourselves. For example the iron content of meat – the main source in our diet – has almost halved over the past years. These nutrients originate in the soil where they are mineralised from soil solids and organic material into an absorbable form. The problem is that our soils are becoming depleted of these substances. More intensive agriculture with its increased use of nitrogen fertiliser at the expense of farmyard manure and the reduced use of diverse rotations have decreased the rate at which minerals are released from the soil, so the old farmers from Clatt were right when they suggested that nitrogen fertilisers 'suck the soil'. Trace minerals are also supplied to the soil as contaminants in some fertilisers and in atmospheric pollution. As we clean up, these sources disappear.

The diet of substantial areas of the southern hemisphere has deficiencies in iron, iodine or vitamin A which are linked to soil degradation. In the northern hemisphere there is a particular problem with selenium deficiency in wheat, which can reduce fertility and immune function and increase the risk of cancers. Selenium can be added to fertilisers, or – an option more palatable to organic farmers – crop varieties could be developed to accumulate selenium. Nevertheless, dietary supplements may still be required to maintain our selenium levels. Soils have even been consumed directly for cures. For example, they are eaten from the sacred places of the Native Americans.

Deficiencies of soil elements also give rise to problems specific to organic farming. Nitrogen capture in legumes depends heavily on a supply of phosphorus and sulphur in the soil. Most soils in the USA and Europe have good levels of these minerals because they were applied regularly in fertilisers. However, these applications usually cease in organic production so levels are progressively reduced and could eventually limit the amount of nitrogen captured by future crops. Phosphorus levels can be difficult to maintain as organic regulations demand use of rock phosphate, which is released very slowly in the soil. Phosphorus recycling is becoming ever more important as the world's supplies of phosphorus are running out.

Unlike nitrogen, minerals cannot be manufactured by burning fossil fuels.

Another reason why organic food is favoured over conventionally produced food is that it contains fewer synthetic pesticide residues and additives. Most certifications for organic farming allow some chemicals and pesticides to be used. Some of these, such as copper sulphate, Bordeaux mixture, rotenone and nicotine, can be toxic. Nevertheless, the overall verdict is that organic food is generally safe. To give a sense of perspective on pesticides, Anthony Trewavas[6] reminded us that plants create within themselves many chemicals that function to kill or deter insect pests. These 'natural pesticides' can prove equally as damaging as synthetic pesticides and our daily consumption of them is far higher than of synthetic pesticides.

Organic farming is usually considered to improve the condition of the soil. In the past few years, I have visited several farms, organic and non-organic, all over Scotland, teaching about the importance of soil structure and soil management for sustaining crop growth and reducing the impacts of climate change. Soil structure was consistently better in the organic farms because they contained both animals and crops, so that organic matter was being added regularly. Moreover, the farms were not managed intensively. The diverse rotations with grass breaks were good for the soil. As in human life, a change is as good as a rest. Just as doing a variety of activities and eating a varied diet rejuvenates us and promotes health and vigour, so too it does for the soil. In an even broader context, use of a variety of crops and animals resembles the rich variety of race and interests required for healthy communities to thrive.

Organic farming improves soil quality by stimulating microbial activity and porosity. The improvement in porosity increases soil water storage and the entry of rainfall into the soil so that runoff and erosion decrease. However, organic farming depends heavily on tillage to bury organic manures and to spread the nitrogen stored in the roots of legumes through the soil. The compaction produced during tillage can cancel out any improvements in quality. For example, when growing organic carrots, mechanical weeding is needed frequently because chemicals cannot be used. This involves many passes over the soil by the tractor and the wheels of the weeder, causing widespread compaction damage. In most situations, though, soil quality under organic farming improves and may help the land to adapt to climate change by making the soil more resistant to damage by storms and heatwaves.

Another way in which organic farming helps to combat climate change is in the avoidance of the carbon dioxide emissions produced during nitrogen fertiliser manufacture. Otherwise, organically farmed land produces the same types of greenhouse gases as conventional farming – nitrous oxide, methane and carbon dioxide. Some of the nitrogen stored during the growth of grass and legumes is lost as greenhouse gases and nitrate leaching after the crops are ploughed in and the first arable crop is sown. Substantial losses of nitrous oxide from the soil at this stage could wipe out much of the saving of carbon dioxide from avoiding nitrogen fertiliser use. Organic farming can be made more climate-friendly by including biofuel production. For example, clover, grass, straw and manure can be fermented together to produce methane gas with a by-product of nitrogen-rich manure.

The lack of mineral nitrogen fertiliser in organic arable cropping is not fully compensated for by adding animal manures, so that soil fertility and crop yields decrease very quickly during two or three seasons. Arable crop yields are also often reduced by increased numbers of weeds, pests and diseases. Average yields per area of land are lower than with conventional farming so that your organic steak ends up with a similar carbon footprint to a conventionally produced steak.

The farmers at Clatt view the widespread adoption of organic farming as an option that could 'land us in a terrible mess' because the spread of weeds and diseases and the low yields could result in food shortages. If Europe were to increase its small fraction of organically farmed land, then we would need to import more food and thereby effectively export the adverse effect on the environment to poorer countries. For organic agriculture to be a sustainable option in the developed world under current consumption trends, the output per hectare of land needs to increase.

Globally, in 2006, organic farming occupied only 0.65% of agricultural land or 30.4 million hectares. The future of organic farming in Europe may depend on market forces and agricultural support. Although widespread use of organic farming might possibly feed most of the population in 2050, this would only happen if we ate much less meat and fewer animal-derived products. If not, more land would have to be taken over from other uses, such as forestry, which would not be sustainable due to increased environmental losses of carbon and biodiversity. Nevertheless, there is a strong niche market for organic food. Organic farmers are traditionally seen as being local, small-scale producers. As the demand for organic foods has increased, though, high-volume sales through supermarkets have become common. Organic farms sometimes become large-scale operations. I visited a research station in Denmark several years ago and I was surprised to see big tractors working in large fields. The spirit of the place was more like that of an intensive conventional farm. Fields can become too big for good rotation of grazing and for prevention of erosion. As a result, there have been moves towards 'sustainable' organic foods which are less industrialised in their production.

Organic farming can increase food security in Africa by producing increased yields for small-scale farmers without the need for inputs of fertilisers and pesticides. In the long term this should help to reduce future dependence on food imports in fluctuating world markets. However, the successful application of organic farming demands a lot of knowledge, with intensive training and continuous access to information and advisory services. Many smallholder farmers do not have education or training and cannot easily get access to information and knowledge. Promisingly, aid agencies are beginning to provide development aid for training in organic farming.

To me the best thing about organic farming is not the principles or the regulations but the mind-set of the organic farmers and producers, which is driven by their connection to the land and to the people. They are extremely motivated and committed to a high standard of management. Their priority is caring for the soil, for the animals, for the people and for the environment, rather than crop productivity. Such farmers perhaps see their farm as an organic whole. This holism extends beyond the farm to the workers making the food products. Joachim Weckmann[7] describes an organic bakery in Germany that is based on a system called Corporate Social Responsibility where the purpose of the company is not to achieve the maximum return for its shareholders, but to produce good products and to contribute a positive benefit to society. The company has a fixed target of providing donations of 10% of net profit to support development initiatives and social engagement of employees in the areas of mentoring, information and training. It also claims to process 100% ecologically certified raw materials into 100% ecological products. This is an exemplary fair trade model which appears to achieve organic farming's goals of ecological, social and economic sustainability. It is a good example of holism extending beyond the farm to interconnect the farmer, the processor and the consumer. Buy organic when you can – especially from those working locally and employing local processors.

Application of the biological approaches of organic farming to conventional agriculture is likely to become more important as fossil fuel supplies dwindle. Many conventional farmers are just as aware as organic farmers of the need to conserve the environment and would be happy to increase their role in this, provided that they can still make a living. The biological approaches with greatest benefits include the use of rotations with nitrogen-storing legumes and of varieties of animals and crops that give good natural resistance to diseases and thus reduce the need for pesticides and medicines. Low-input farming is one such mixture of organic and conventional farming – called Integrated Farm Management (IFM) in the UK – and claims to combine beneficial natural processes into modern farming practices using advanced technology. This and other forms of alternative or low-input agriculture, termed 'ecologically sustainable', clearly accord with the principles of deep ecology since they emphasise independence, community, harmony with nature, diversity and restraint.

Marketing the produce from low-input farming suffers because it is not so readily identifiable as healthy and environmentally friendly as organic food. If you can recognise it, then buy it. However, like organic farming, low-input systems are unlikely to be sustainable at a large scale in current economic conditions as they do not yield enough. Other approaches are also needed to maintain our burgeoning population while conserving soil, water and the environment.

Chapter 8

RESTORATION AND CONSERVATION

'Restoring soils to improve food security ...
what other human right deserves higher priority?'
Rattan Lal[1]

My first experience of restoration was with a motorbike, a 1955
Ariel 500cc. I had left home and started work by this time. During
my visits home, when travelling between Clatt and Rhynie, I had
noticed the rear of an old bike in a shed just visible from the road.
In my mind I recalled those heady days of flitting through the
country lanes on the Norton when the weather was mild, the sun
shining and the air rushing at me fresh and sweet. Soon afterwards,
as I was walking past the shed, I got into conversation with the
owner – and you can guess the rest. The bike was mostly complete,
but rough, with all the chrome bits well rusted. It proved an arduous
job right from the start. First, I had to push it the three miles home
to Roadside. I was exhausted by the time I got it there and started
my 'labour of love' as most around me called it. This involved
skinned knuckles, a lot of swearing and sweat, a search for spare
parts and a rapidly depleting bank account, all of which seemed to
occupy most of my spare time. Maybe I should have brought home
a woman instead.

The biggest job was the beautifying. I dismantled the bike and
sent off many of the larger parts to be re-chromed and repainted. I
was disappointed when the paintwork came back. The bright red
bore little relationship to the original deep crimson lake. My first
lesson was that nothing is perfect. I had sent the engine down south
to be restored by a specialist. When I got this back, after 6 months,
I soon had the bike back together and it looked great, even though
it was the wrong colour. Now at last I could bring it back to life. But

I could hardly move the engine with the kick-starter. The engine restorer had told me that 'she would be tight'. I tried everything to turn the engine. I could do it a little bit at a time, but it needed real power to turn it fast enough to start. Eventually I got one of my mates to give me a tow with his tractor. He pulled me unceremoniously through Clatt with the wheels of the bike turning the engine through the drive-chain and gearbox. This was nerve-wracking as the tractor had a back loader with bits of muck dripping from it onto the road and onto my precious bike. Also I knew I had to be able to control the bike once it fired up, which it did eventually. All of this was not as I had imagined and was deflating for the ego.

I soon had the bike running well, though mid-50s' British bikes were never smooth and reliable even when new. It proved daunting to drive in modern traffic because the brakes were not very powerful and on long journeys the vibration was crippling. But the loving relationship held and I was sorry to see it go when I eventually sold it at a knock-down price to a dealer in south-west England. However, the knowledge, wisdom and patience that I gained have stayed with me ever since.

The motorbike and its restoration depended almost entirely on the use of metal ores and fossil fuels – oil in particular, a commodity about to become scarce and expensive. In looking for sustainable agricultural systems beyond organic it is necessary to have options that are less oil-dependent both to restore damaged land and to conserve it. Like the motorbike, restoring an ecosystem and the soil below it demands effort and is a 'labour of love'. A successfully restored ecosystem gains beauty from the new vegetation or the crop which protects and hides the soil. Indigenous peoples believe that they have an obligation to keep the planet alive, and they enter into a reciprocal relationship with the environment. Leslie Gray[2] believes that this is why restoration and conservation are important to them. Tending the world also empowers them and gives energy and health.

Land is not naturally agricultural. In most areas it was originally forest, scrub or rough grazing and was brought into an agriculturally productive condition by a process of reclamation. Many reclamation methods resemble restoration methods. The dictionary definition of reclamation is 'to call back, to win from evil, wildness or waste'. I'm not suggesting that our naturally vegetated soils are evil, but they are certainly wild and land reclamation is essential for the survival of our ever-expanding population.

Deforestation of tropical regions for land reclamation has gained bad publicity. However, the beneficial effects of reclamation can be observed everywhere. For example, in Clatt, the neat, productive fields surrounded by woodland delight the eye, but it's easy to forget the huge efforts put into reclaiming this land from moor and bog by our forebears. From the 1745 rebellion onwards, the peasant folk of Scotland turned moors and bogs into 'smiling fields'. This involved trenching, draining and enclosing – work that had to be done from early till late in all weathers. Allan[3] describes how a trench about 1 metre wide and 30 cm deep was made by digging out sods of heather or bog, which were carried off to one side in a barrow. The sods were then removed from an adjacent 1-metre-wide strip and put face down in the first trench. Next, using a pick, the stones were loosened and removed from the bottom of the second trench to a depth of 30 cm, and the soil was then lifted with a spade onto the sods in the first trench. This procedure was then repeated for a third and subsequent strips until the whole field was turned over. The roots of the broom bushes and other native vegetation were also removed and spread elsewhere to rot down. All this labour was done by hand. Only later was the decaying sod ploughed up prior to planting potatoes or turnips. Allan describes this as 'waging a hard warfare with Nature'. The result of this 'warfare' or struggle is the wonderful pattern of fields of cereal crops and grassland in a typical Scottish landscape. In upland areas, such fields can lie alongside native heather moor and forest as a testament to the lasting success of reclamation (Figure 15).

Restoration attempts to achieve quality and beauty as part of a vision for a healthier world, working with the regenerative powers of nature to give the land back its productivity. The restoration of soils which have been abused as a result of compaction or erosion often uses similar techniques to those of reclamation. Whatever methods are used, they should preserve porosity, allowing easy movement of rainwater into the soil and water storage. Soil organisms need to be kept active so that they can break down organic matter and release the nutrients stored in the soil, while retaining recalcitrant carbon. The loosening and mixing action of ploughing achieves this well, when done in the correct conditions. Ploughing is like turning over a new leaf, or, more accurately, an old leaf. Well-ploughed land is level, with broken-up aggregates and most of the surface vegetation buried so that it looks good. Such seeking for beauty usually pays off because the land can be quickly made into a productive seedbed. However, burial of the surface material hides weed seeds which may germinate later. It's like in our own lives – sometimes burying the past and making a new beginning is not completely effective: intrusive memories may return unbidden for several years, as in post-traumatic stress disorder.

Good restoration techniques all tend to dry out the soil and promote the development of soil structure. Examples of this are subsoiling, which loosens deep compaction, and drainage, which improves wet, low-lying or peaty soils. Nevertheless, improving or conserving the soil is best done either by living organisms such as earthworms or by plant roots, which are excellent natural creators of macropores that are stable and interconnected, enabling them to maintain the soil's life-support functions. Where there is a risk of erosion, maintaining plant cover at the surface and organic matter is important to provide protection and, in drier climates, to increase green-water storage and crop yields. In these ways the soil is rejuvenated, especially under a mixed crop rotation when slow-releasing minerals like lime and phosphates are added. Many will remember

Detroit, or Motown, the home of so much well-loved soul music. The motor industry has now largely deserted Detroit and in its place urban agriculture is being practised on land damaged by industrialisation during the 20th century. There are large areas of compacted, vacant land left behind by the car industry that are being restored for cropping by loosening the soil, removing debris and contaminants, and using plants to absorb toxic chemicals.

Structural damage to the soil can look worse than it actually is. For example, pigs can churn up the soil into a wet mess. I was given the unenviable task of finding out how much damage this caused to soil structure. To make sure that conditions were poor, I ended up doing this in February in the freezing cold and wet, handling electric fences and trying to avoid upsetting pregnant sows or new mothers. I discovered no evidence of lasting damage to the structure and much of the compaction in the topsoil could be loosened by ploughing. The main problem was that the soil was broken up and exposed to the rains so that on sloping, sandy sites water erosion was occurring. If soil is kept in reasonable condition, its structure is remarkably resilient. I did this work on farms in south-east England, so I would go down the night before, start early and work until about 3 pm when a colleague would rush me to a station to get the last train up to Edinburgh. I would step onto this train, covered in mud, freezing cold and smelly. I was also hungry so I went along to the dining car in First Class and usually managed to get a table to myself. It was a curious feeling sitting there surrounded by suited executives and comfortably-off retired couples, a world so different from the one that I had just left.

Soil compaction damage is the result of stress which, in extreme cases, can be considered as violence to the soil. In restoring compacted soils, it is helpful to bear in mind that violence usually does not overcome the results of violence. The soil can be cultivated using large, heavy, rotating metal spikes, called power-driven tine cultivators, which effectively exert more violence to try to bash the soil back into shape. These often cause more compaction and destroy the natural soil structure. Instead it is preferable to tease apart the soil by carefully targeted loosening. I sometimes think that there is a parallel between tillage and counselling, which frees up hard, inflexible areas in our minds to allow creative, positive thinking and awareness of our self-integrity. We need to loosen soils when conditions are just right, neither too wet nor too dry. In a similar way, cultivating good behaviour, for example with children, is best done when conditions are good, usually when everybody is calm. In the case of excessive cultivation, the soil at the surface tends to slump together to form a layer or cap that prevents the crop from emerging from the soil, thereby restricting growth. Although this surface layer goes hard on drying, the integrity of soil as a whole is maintained, affording protection. This type of non-cooperative reaction to poor soil management reminds me of the successful use of 'constructive non-violence' by people opposed to oppressive regimes – such as those that brought about the removal of the Iron Curtain or the end of apartheid in South Africa.

In north-west Scotland and in the Hebrides, crofters are starting to restore to agriculture the peaty soils abandoned by their predecessors over the last 50 years. The landscape is beginning to look alive once more. Restoration by drainage, tillage and liming needs to be guided by the depth of the peat, by the vegetation cover and by the drainage status, all of which can vary widely over short distances. As with organic farming, a mixed farming system is used and animals are involved. It's hard going, often in poor weather conditions. On top of this, the over-zealous enforcement on crofts of environmental pollution regulations that are usually designed for larger-scale enterprises often causes despair and abandonment of efforts. Restoration should aim for the best for the soil, the landscape and the people, dependent on local conditions.

The people involved in land restoration often include volunteers, who become aware of the abuse that the soil has suffered and who bond with the land as it heals. Elan Shapiro[4] considers that this creates a connection at a deep level and engenders feelings of pleasure and release, opening up channels to allow us to unload our sense of guilt, shame, grief, loneliness or despair. She considers that such work, done with others, allows us to embrace both our fragments and our integrity and helps us 'to reweave the tattered fabric of our souls'. The healing of the people and the planet belong together.

Conservation is considered as being economical or sparing with a resource such as fuel or water. Our parents or grandparents naturally conserved things, turning off lights and heaters without thinking, perhaps prompted by unconscious memories of scarcity, expense or value. This type of conservation is relatively easy, though many of us choose not to practise it. Indeed, Thom Hartmann[5] believes that living frugally leads to a feeling of accomplishment and independence.

Conservation applies to the preservation of food as well. At Roadside, lots of fruit was conserved with sugar as jam. I spent many happy hours picking wild raspberries in the hedgerows around Clatt. Yet I left many behind. In a similar way about one third of the food that we produce never actually gets out of the field and rots there, particularly in developing countries. And, despite our fridges and freezers, we still manage to throw away about one third of the food we buy.

Conservation also means preservation or keeping whole. After restoring my motorbike, it soon began to leak oil, to get dirty, to lose its shine and to rattle as bits slackened. I had to carry on working to preserve it and to conserve it. Conservation involves cycles of regeneration, giving and returning. The soil is a finite resource, so it needs to be kept in good condition to minimise the chances of degradation or loss by preserving the usefulness of the water, the organic matter, nutrients and creatures in the soil. The importance of conservation was realised by Vernon Carter and Tom Dale[6] who observed that 'no ancient civilisation had an effective conservation policy'.

The farmer is central to the conservation process. Wendell Berry[7] considers that farmers not only produce food, but also, by default, conserve soil, water, wildlife, open space and scenery. Conservation also means asking of soils, agriculture, people, not what we can get out of them, but what we can give back. A farmer or soil scientist might ask, 'What can I learn from this agricultural system that will allow me to manage it for maximum yield?' Instead, we need to ask, 'What can I learn from this agricultural system that will enable me to serve it, or con-serve it, better?'

The primary aim of management should be towards sustainability by conserving the soil. This leads to a 'win-win' situation, because improving soil quality increases carbon storage, reduces runoff and, as a result, improves soil productivity. The older farmers at Clatt have developed an instinctive way of conserving the soil. To improve its quality, they use rotations with break crops and grass, regular farmyard manure applications and periods of sheep grazing. Rotations effectively allow for periods when the soil can rest and recuperate. These techniques, developed over the years, are based on the wisdom of our ancestors and work well, perhaps explaining why farmers are unwilling to change quickly to adopt new management techniques. Wendell Berry[7] considers that conservation involves more than just changes in management; it also involves the heart of the person managing the land. If they love the soil, they will save it.

Land prone to erosion is often managed by conservation agriculture, the main principles of which are to prolong permanent plant or residue cover, to use diverse crop rotations, to minimise soil disturbance and to leave crop residues at the surface. An important component of this is the use of conservation tillage (either no-tillage or minimum tillage) instead of ploughing. In no-tillage the seeds are drilled directly into the soil under the residues of the previous crop, which protect the seedlings (Figure 16). Minimum tillage uses some shallow fixed-blade or disc cultivation before the sowing of the seeds. Conservation tillage uses less machinery and fuel than normal ploughing but depends heavily on weed control by chemicals and demands high standards of management. It is extensively used in Brazil and Australia, but its continued use is threatened by a build-up of soil compaction and resistance by the weeds to the chemicals used for their control.

Perennial crops perhaps offer the ultimate potential for large-scale conservation agriculture. Unlike most crops, the land needs no regular tillage because crops do not need to be re-sown every year. Stan Cox[8] claims that a field of perennial grain crops would provide food while conserving soils and water and reducing inputs of fertilisers, chemicals and fuel. They could eventually replace the huge areas of annual cereals currently being grown. Because they have a deep, extensive root system, like native perennials, they are well-suited to marginal land subject to degradation, access more water than annual crops and perhaps offer benefits of carbon storage. Although perennial crops like alfalfa and clover are well established, an extensive, long-term plant-breeding programme is necessary to produce cereal varieties which will yield enough grain to be economic.

Smaller-scale conservation agriculture methods which may help develop sustainability include the revival of old farming systems where land was shared and farmed in small parcels such as crofts or allotments and which help to increase local food crops. More novel systems include agroforestry, where trees and food crops are grown together. For example, in India wheat is grown for food amongst poplar trees which capture carbon in their wood, roots and the soil. Crops can also be grown by aquaculture, using brackish water or seawater.

The ultimate aim of agricultural sustainability would be to develop less intensive, small-scale farming systems very similar to natural ecosystems and matching local conditions. Multiple cropping in the same field or garden is common in some smallholder farming systems. This encourages many different organisms to flourish, is highly productive and gives good insurance against the risks of erosion, weather, pests and disease. However, it demands skills and knowledge which are hard to acquire.

One system that concentrates on producing soil of high quality, leading to good productivity, is biointensive farming, also known as mini farming. This is based on biodynamic organic farming and intensive raised beds. It builds on the wisdom of our ancestors, having been in use for 4000 years in Asia. It uses little machinery or fossil fuel inputs, but is labour intensive. Soil is double dug to twice normal spade depth and organic matter is added as compost. Crops are planted close to each other, often in combinations of different types which grow well together. A relatively small area can support someone on a vegan diet and provide a small income. It is particularly relevant to small farmers, often women, in countries where land and resources are scarce and it could help transform the lives of the rural poor. Good training is required so that all aspects of the method are followed, otherwise the system can fail and cause soil degradation. It has to be adopted with a spirit of full commitment.

Another form of crop cultivation based on ancient practice is permaculture, a system of permanent agriculture and permanent culture which goes beyond biodynamics or biointensive farming. It has its roots in the highly efficient 'permanent farming' practised for many centuries in China, Korea and Japan. Soils professor F. King[9] describes how every square metre of land and litre of fresh water were used and how everything from human waste to canal dredgings was recycled. This was combined with an intense economy of practices by the people in their efforts and lifestyle. Permaculture has an ethical component in that it embraces care of the Earth, care of the people and fair sharing[10]. It relies on agroforestry and perennial food crops to give self-sufficiency with minimal amounts of energy use, and recycles most wastes, including human wastes. Although it sounds good, Joel Salatin[11] thinks that permaculture has limitations in practice because much landscape manipulation is needed to form ponds and wetlands. This, in combination with current government wetland regulations, would limit its application in drier areas.

These methods all help to localise food production. Extending this to towns and cities will require the development of urban agriculture and horticulture. Urban food forests will also most likely be important.

Although restoration, local small-scale conservation farming and organic agriculture are crucial to tackling rural poverty and contributing to food security, large-scale industrial farmers are faced with current demands to produce more. According to the United Nations, world food production needs to increase by about 40% by 2030 in order to meet growing demand. This has to occur on less land with less water and energy use. Reconciling this demand with the need to make a profit, conserve the environment and mitigate climate change is a difficult task.

Land management and land use are the most important components of successful agriculture. If the soil and climate are suitable and enough resources – especially water and fertiliser – are thrown at the land, it will be productive. This can be observed in many temperate areas. For example, on the sides of Tap O'Noth, near Clatt, forestry, intensive grazing and heather moor all exist within a few metres of each other (Figure 15). However, this comes at a price. Jonathan Foley[12] considers that changes in land use by expanding the areas under cropping, pasture, plantations and towns have caused huge increases in energy, water and fertiliser consumption and biodiversity losses, all of which have a greater destructive effect on the planet than climate change and is ultimately not sustainable.

In Clatt, the farmers, like many others, have depended on increasing production either by expanding their land area or by farming their existing land more intensively to maintain profits. Intensification means getting more product per area of land and is achieved by use of improved breeds of cereals, chemicals to control pests and simplifying crop rotations so that cereals dominate. In the 18th century, the fertile land around Clatt would have looked quite different from today. Flax was widely grown for making linen yarn. Special spinning wheels were used for this, examples of which were made in the village. Perhaps unknown to the spinners, some of the linen garments were exported to Jamaica to be worn by slaves producing sugar and rum for consumption in Aberdeen, an early indicator of a link between fibre production and injustice.

Changes in land use respond to market demands. For example, when I was a student, a large part of the grouse moor of Suie Hill in Clatt was reclaimed for grassland and forest by ploughing, liming and fertilising the heather. As kids we spent time there camping, exploring and, in later years, flapping flags to dislodge the unfortunate grouse ahead of the guns of the gentry. Part of the area was known as the 'berry face' where villagers went regularly to pick blueberries and cranberries, which gave a hint of the potential fertility of the land.

In time, the laird's influence in Clatt slipped away as he sold farms and the number of incomers grew. In 1989 the entire Knockespoch estate was sold to an outside buyer rather than to the tenants as this realised more money. This led to a time of uncertainty and a feeling of powerlessness for the tenants. However, the new laird from London had insufficient funds to develop the estate, so it went back on the market. This time the tenant farmers managed to buy their farms as a consortium and finally owned the land that they and their forebears had worked on year after year. Since then there has been some succession of farmers, but many of their children – my schoolmates – had no real interest in farming. It is not clear where the next generation of farmers will come from.

I have spoken a lot about the love of the land felt by farmers, but as farming intensifies it becomes ever more demanding. Larger machines have replaced farm labourers so that many farmers now work on their own. Also, looking after livestock and arable land is a 24-hour, 7-days-a-week occupation. It may sound romantic to work on the land but the farmers that I know work very hard indeed and struggle to keep up with paperwork and regulations. This is in contrast to the 19th century in Aberdeenshire, when the basic task of ploughing was not so much a chore as a nurtured art fostered by ploughing matches, which were very popular. Allan[3] recalls how 19,000 farmers and farm servants turned up to a championship match near Inverurie and enthused over uniform straight furrows, admiring how others strove to achieve beauty and quality with the soil. This is a spiritual value, part of the making of the land for the crops which Wendell Berry[7] calls the 'giving of love to the work of the hands'. Allan[3] considers that skilled work well done engendered a true sense of happiness and pride which he thought was being lost. He realised that in 20th-century Scotland men only turned out in such numbers to watch football matches.

I recently went back to Clatt, to a field near Roadside where a friend let me plough Insch series again. This was good, though getting to grips with driving a modern tractor with umpteen gears and a plethora of controls was more difficult than the actual ploughing. Much of the farm had been cropped continuously to cereals for years by the previous farmer, with little addition of organic matter. The field had been under snow for a long time during the previous winter. As a result, when I inspected the soil using the spade test, it was compact, cloddy and rather grey (Figure 17). I was somewhat disappointed at the lack of beauty. Even soils of high natural fertility such as Insch series demand drainage, the addition of organic material and care to prevent damage, as climate change and associated weather extremes make them more vulnerable. Although regular applications of farmyard manure appear at first sight to sustain cereal growth year after year, mineral fertiliser is still necessary to maintain high yields. So nitrous oxide and carbon dioxide are still steaming off the soils to fuel climate change, and nitrate continues to leach from the soil into the cheerful Gadie Burn.

All over the world, technologies of nitrogen fertilisers, irrigation water and pesticides continue to be abused by excessive use in intensive agriculture, all of which can decrease soil quality. For example, in New Zealand, where the export of dairy products is vital to the economy, the dairy systems are intensive and expanding. I recently spent some time there measuring soil quality and nitrous oxide emission in pasture under dairy production. As I worked in the field, my wife waited in the car and watched the cows coming out from milking. So many walked past that she eventually thought they were going round in circles! Dairy production is becoming increasingly unpopular with the local population because the augmented use of irrigation water is drying out the rivers, and the removal of tree shelter-belts to accommodate the very wide-boom irrigators is making soil erosion more likely. In addition, the increased discharge of effluent high in nitrogen is threatening water quality and there are concerns for the welfare of the animals. Soil structure can be damaged by compaction from animal hooves and wheel traffic (Figure 10) which restricts the growth of the pasture shoots and roots due to low macroporosity, reduced biological diversity and activity, and restricted nutrient uptake. If extra mineral nitrogen fertiliser is applied to compensate, only part is used by the grass with the rest lost to the environment.

In the future, while fuel for the creation of nitrogen fertiliser still remains affordable, intensification needs to become more ecological. Not only does productivity for a given area of land need to increase substantially to avoid the takeover of yet more land for agriculture but the number of inputs per product should be reduced to minimise any environmental impacts. Eco-efficient farming systems will have to be highly productive, relying on clean energy sources and using environmentally-favourable industrial processes and sustainable agricultural practices. Rattan Lal[13] shows that these practices should conserve soil quality, increase carbon storage in the root zone and improve natural water quality. Mineral nitrogen input will have to be reduced by more efficient use, recycling and exploitation of natural sources.

Innovative solutions are required. An important move towards improving the sustainability of New Zealand dairy systems is to increase eco-efficiency by maintaining high soil and pasture quality. Graham Shepherd[14] uses visual methods of crop and soil observation to adjust management to improve the contribution of soil biology. Improved soil management involves some tillage where necessary to ensure good soil structure and aeration, the choice of appropriate mixtures of grass, herbs and clover in the pasture, the use of simple pasture-reseeding techniques and the application of biologically-friendly fertilisers that promote carbon sequestration, nutrient turnover and cycling, and the slow release of nitrogen to the pasture. These combine to encourage earthworm and microbial activity in the soil and allow good pasture production with low nitrogen fertiliser use. Healthy pasture supports healthy animals, and greenhouse gas emissions are reduced relatively cheaply. Conservation shifts the emphasis from agribusiness to husbandry[15] where the farmer aims for a reasonable return rather than high productivity. Perhaps the best standard of agricultural performance is not speed, efficiency and productivity, but the health of the ecosystem, the farm, the soil and the human community.

A similar approach, which emphasises 'soil first', is taken by the New Zealand dairy farmer Max Purnell[16]. He regards the soil as a rechargeable battery that stores solar power. He considers that the best way to grow his business is to make the battery bigger by increasing the depth and the function of the soil – in other words, growing soil. To increase the speed of recharge, he recommends encouraging root-zone fungi by growing pasture with different types of grasses, herbs and clovers to their full life cycle.

A pasture-based system of biological agriculture labelled by some as 'beyond organic' was developed by the family of Joel Salatin on his Polyface farms – farms with many faces[11]. The land on their original farm in Virginia, USA was heavily eroded with many areas of exposed bare rock, making the surface appear like pockmarked skin. By applying compost, rotating cattle and running poultry, these sores developed scabs of soil and eventually healed as the soil reformed over a 50-year period. Enough soil was created to refill many gullies. This farming stimulates soil biology and carbon sequestration. Plants and animals are allowed habitats which suit their physiology in systems which aim to mimic natural biological conditions. Economically viable agricultural enterprises are developed, with emphasis on healing not only the land, but also the food, the economy and the culture. I see these systems as engaging and continually renewing the spirit within the farmer, though Salatin cheerfully calls himself a lunatic farmer! Nevertheless, he describes his spirit soaring as he steps out into the 'fresh-scented morning air' of his farm. The food from Polyface farms is tasty and is claimed to be nutritious enough to keep people out of hospital. But what most impresses me about this 'lunatic farmer' is that he refuses to sell his produce by mail order and recommends travelling to the nearest Polyface farm instead. That is true commitment to local marketing and sustainability.

These and other innovative solutions to food security need to be communicated to farmers in a way that is readily understood. Intensive farming depends heavily on the use of machinery – and farmers just love machinery. I recently attended a vintage tractor rally in Clatt where we spent a day and a half poring over old tractors and reminiscing. I was the only person to mention the soil over which the tractors passed. Similarly, at agricultural shows the machinery stands are always busy. Aligning novel techniques with machinery use and the potential to save on fuel and labour may help farmers to restore attention to their soils and crops.

When advisors or scientists speak to farmers about new crop production methods, the farmers are wary. Their appetite for change and development varies greatly. With the internet, they have access to a great amount of knowledge but often find it difficult to identify what is appropriate and develop wisdom for its correct application. I empathise with these farmers. It reminds me of why I bought and restored the Ariel motorbike. I tackled it because the attraction of the novel features of modern motorcycles such as greater efficiency, power and speed had worn off and I sought the security of something well-designed and simple, something that was good enough for my elders and that I remembered made me feel good and carefree when I was younger.

Advice is often provided free to farmers by suppliers of fertilisers, pesticides or machinery and thus may be biased. Knowledge transfer is often hampered by a lack of consistent, unbiased advice. The use of technology without wisdom is considered by Rattan Lal[17] to be one of the great blunders of humanity. Yet there are examples of wise use of technology hidden within. In my community in north-east Scotland, a contractor told me that he determines the optimum depth for subsoiling by first crossing the field once with the subsoiler at full depth. He then feels down the subsoiler legs to identify the zone with the warmest metal. This corresponds to the average depth of the compact layer and he then adjusts the subsoiler to penetrate to just below this depth. This information appeared after the consumption of one or two whiskies – a case of one spirit enabling the other!

Although solutions for better farming are mostly farmer-centred, farmers and advisors should get together to develop new production methods from innovative ideas produced by scientists. These new agricultural methods need to integrate easily with the old ones and be developed or adapted to local circumstances, especially where workers are poor, partially skilled or partially educated. Much depends on learning by example. Farmers look over the hedge and, if they see something working, they often adopt it. Alastair McIntosh[18] believes that to carry over wisdom may require a revitalised connection between the elders and a younger generation. Development of wisdom was perceived by Fritz Schumacher[19] as the orientation of science and technology towards the holistic, the gentle, the non-violent, the elegant and the beautiful.

Fostering the desire for sustainability and change perhaps needs a more direct engagement of the spirit within to help farmers to share and develop further the wisdom they have gained from their love of the land and its husbandry. I discovered this when I demonstrated the spade test to farmer groups. The careful, respectful handling of the soil released a flow of ideas and experiences which in turn encouraged others to talk. This is growth from within and is similar to the growth and spread of plants which have rhizomes or corms. Rhizomes are thick stems like roots that spread out sideways in the soil and send up shoots at random. Orchids, bananas and some grass weeds grow this way. Cutting up the rhizomes causes them to spread rather than to be destroyed. This is a better model of how information spreads than that of a tree, where it filters from above to the grass roots. Dolores LeChapelle and Julien Puzey[20] suggest that growth from below spreads methods for healing the destruction of the earth. Each community or village can develop within its own area, protesting and growing, bringing an infection of restoration and conservation. This would help to expand and remould the knowledge that farmers already have. I think that there is a parallel with the restoration of our individual health and confidence. We can reduce the intensity of our lives by relying less on chemicals to keep us sane and taking a more holistic approach, sharing with others, being rational in trying to identify the roots of our problems and allowing ourselves to find our own solutions.

The development of more sustainable practices will also require changes in our diet. The farmers in Clatt grow many cereals, mostly barley, which are either fed to cattle or pigs or used to make malt in whisky production. These are great for that most Scottish of meals, the Burns supper, but not so useful for everyday consumption. The only crops grown directly for food are small areas of wheat, oats and potatoes. World meat production is predicted to double between 2000 and 2050. Grain is fed to animals to grow much of the world's meat. This is inefficient. It takes, on average, 6.5 kg of grain, 36 kg of roughage (hay, grass and silage) and 15 000 litres of water to make 1 kg of beef. It is more efficient for us to eat the grain rather than feeding it to animals to make meat. Such a transition may help organic farming and conservation agriculture to feed more and more people. To achieve this, whole cultures may need to adjust to eating different types of food. This requires a change from within, brought about by the development of a spirit of conservation.

Agricultural management is clearly crucial to our future, but not all land needs to be used for agriculture, especially marginal land in developing countries which is difficult to improve. Land users could take advantage of abundant sunlight, wind and space by developing solar and wind energy and tourism. Providing rural people and poor farmers with the opportunity to earn sustainable, stable livelihoods will help to conserve the planet's biodiversity and to bring beauty to the landscape and the soil, justice to the people and reduction in poverty through improved diet.

The methods of increasing production by improved efficiency, fostering soil restoration and conservation and soil biology, though very important, are unlikely to be enough in a resource-poor future. Most of our food production technology – even organic – is still energy intensive. It takes 10 calories of petroleum to make each calorie of food we eat. Another myth has been exposed: that we can be saved by science and technology. We all need a fundamental change in mindset. For this we need to engage the spirit within.

From oil to soil[21]

Free oil
Drilling down
and gushing forth
Stored sunlight from ancient earth
Who needs soil when you've got all that oil
Flying free, driving on, population exploding in oil
Artificial fertilisers, plastics, synthetics, tar, cement and spirits abound
Peak oil, who cares when there's still half left below, deep oil
Gas fracking*, open cast, tarry sand, rip it out as it lasts
It's OK to plunder the ancient landscapes below
Who needs soil when you've got tough oil**
Not our problem, business as usual
Technology always advances
Can we make it
digging down
with soil?
Oil free

---

*   Gas fracking is a controversial process of extracting gas and oil from shale by hydraulic fracturing of rocks. The chemicals used can pollute water supplies underground and gas may be released into drinking water supplies.

** Tough oil describes much of our remaining oil which is difficult, risky and damaging to extract.

Chapter 9

A PEACEFUL STRUGGLE

The difference between what we do and what we are capable
of doing would suffice to solve most of the world's problems
Mohandas Gandhi

The soil scientists Vernon Gill Carter and Tom Dale recognised back
in the 1970s the need to 'adjust the population and the standard of
living to usable resources'. They realised that our crisis is one of
overconsumption and an unjust distribution of consumption. The
growing demand for food by the world's swelling population and its
changing diet is very unlikely to be met in the near future because
we simply haven't enough fossil fuel, water, minerals and soil to sustain us. In the northern hemisphere, we all need to consume less,
waste less and recycle more of everything. In the southern hemisphere, in areas of constraints of poverty and education, consideration needs to be given to reducing the number of mouths to feed by
limiting family sizes without coercion. Improved educational and
employment opportunities for women, in particular, would help
achieve this, as it has in countries of the north. The global population could then live within the capacity of the land and diminishing
resources – particularly of fossil fuels. Carter and Dale[1] also predicted that a failure to balance the number of people with available
resources would require us to adapt to a lower standard of living.
We need to become aware of the seriousness of the situation we are
facing. The consequences of ignoring the implications of peak oil,
indeed of peak everything, could be a huge failure of a substantial
part of our industrial civilisation, according to Charles Hall and
Kent Klitgard[2]. Almost all transportation – including electric vehicles – depends on oil. The clothes we wear, the furniture and fittings
of the room where you're reading this, and the prescription pills

which keep so many of us going all depend on oil.

According to Thom Hartmann[3], our basic problems lie not in energy but in our view of life which is grounded in our culture. Our world is at risk due to widespread injustice, inequality and disharmony and the only thing that will save us is to change the way we see and understand it. Our world view is shaped by prejudices, core beliefs and apathy ingrained within our 'subsoil', or shadow self, which rule our behaviour. We thus need to become open to a change in awareness within ourselves, which is a spiritual process. George MacLeod[4], the founder of the Iona Community, said, 'The one supreme conviction that I cannot get away from and – without any dramatics – am quite willing to die for, is that only the spiritual can mould any future worth having in the world.'

In order to work towards a sustainable future, we need to stimulate our own spiritual life by accepting the direction of the spirit within nature and by developing our connections with the springs of love deep inside us. The changes at Roadside and Clatt during my lifetime give some clues about how this might be achieved. The hard living conditions at Roadside eventually took their toll on my parents. The house began to deteriorate and to fall apart. The Old Man had no money to rebuild or improve it and he was too old to repair it himself. The last straw came in 1981 when Clatt shop and post office closed, so we at last persuaded him to leave Roadside and to move into a council flat in the nearby town of Huntly, much to my mother's relief. How she loved to run those taps! But the Old Man never liked their new home and died the next year. We were still trying to sell Roadside at the time and, by a remarkable coincidence, I sold it on Easter Sunday, on the third day after his death on Good Friday. It was as if it had been waiting for him to go.

The next occupants of Roadside, attracted by the remote location, improved it by removing ceilings, putting in doors, making a bedroom from the attic that we had never entered and installing some plumbing. But they still struggled to live there because there was no mains water, the fabric of the house was still of 19th-century standard and access by vehicles was difficult. Once again, the community provided support, but by this time improved communications had made people look more to the outside world and less interested in the problems of a newcomer. Today Roadside is empty and neglected.

In the village, two initiatives developed within the community that extended the use of the traditional meeting places of the church and village hall. The numbers attending church had dwindled to such a low level that it closed in 1985. The spiritual heart of the community was lost. No longer would droplets of Clatt water slip between the fingers of the minister onto the foreheads of the newborn of the village. As when the soil becomes depleted of organic matter, much of the glue holding the community together was gone. Most of those still attending church at closure were the elders of the community, mainly farmers.

The church building was important not only because of its beauty and prominence but also because that was where the villagers celebrated marriages and baptisms and conducted funerals, occasions when the common ground of being – the deep Self – was shared. Clatt Kirk (Figure 7) goes back a long way to its founding in the 1st century AD by St Moluag, a Celtic missionary who lived at the same time as St Columba. The Church of Scotland threatened to sell the church and the community was faced with the possibility of a ruin in the middle of the churchyard. There was sufficient interest in saving the building to form a committee which raised enough money, mainly through the National Lottery, to restore it. The Old Man had loved it and, when I was a boy, he wrote:

> *I would hate to leave her*
> *Our little village church*
> *Standing on a grassy mound*
> *She can easily be found*

Looking at the landscape
Her windows are her eyes
Her flock she sees on Sundays
As they come up the aisle

God's house is at their call
All are free to use it
His loved sanctuary of peace
For here, Hope does not cease

Clatt Auld Kirk, as it is called these days, now hosts musical evenings, concerts, lectures and occasional weddings and funerals. It is surrounded by graves containing our ancestors and I think that we somehow tap into their wisdom when we are there. In the Middle Ages, the churchyard was used as a meeting place for markets so it is now returning to that earlier use. The former Church of Scotland parish of Clatt was combined into the larger linked parish of Noth and, at the time of writing, regular services are held in Rhynie, four miles away. The parish celebrated its 25th anniversary in 2010 when the minister was still trying to persuade those in the parishes that were 'untied' back in 1985 to attend at Noth.

The other community initiative was the development and extension of the village hall as a tea room and shop in 1985, to give young people employment and to provide a market for local produce. It was a popular venue for both visitors and locals at weekends. In the early years an elder villager was often available to recall life in the old days, a living connection with the past. The Clatt people proudly showed their history in displays of photographs and plaques around the walls of the hall. The common feature of these photographs is that they show groups of people from 50-100 years ago, small communities linked together by shared schooling, employment or interests – or in some cases just living together in Clatt. We no longer seem to do this. When last were you photographed in a group of more than two or three people other than your own family?

The developments in church and hall are based on buildings, the special meeting places that people remember. I think that people gain comfort from the constancy of place and soil, but neither of these initiatives has run smoothly, as the people who best remember the places move on or lose interest. The tea room is now open only occasionally. The committees that run them struggle to find enough committed people and the members occasionally fall out. In this way Clatt is no different from many other villages where 'community' has lost its meaning of shared living and relates more to a collection of residents who happen to live in the same place or a group brought together by a common interest. The 'communities' drawn together by the events at the hall and the church include tractor enthusiasts, bikers, dog owners and music lovers from all over the north-east of Scotland.

Like many other places in the Western world, 21st-century living in Clatt brings problems of obesity, alcoholism, anxiety and depression, materialism, isolation from neighbours and lack of spiritual awareness. There is also a great dependence on cars, big supermarkets and fossil fuels. We are living in a social recession and we try to cope with the threats of inequality, violence and environmental destruction by use of short-term fixes like retail therapy, overeating and alcohol. The villagers struggle between the pull of the slower pace of the past and the urgent push of the future which crowds them into greater productivity and consumption.

The current changes in society and lifestyle, occurring at an ever-increasing pace, undermine stability and any sense of permanence. Some spiritual input is required to sustain the community, the soil and agriculture, and to give a sense of lasting satisfaction. I found a clue to the form of this input in Roslin, in the Community of the Transfiguration. It took me a long time to even get the name fixed in my head, but it was the name that gave me the clue. The word 'transfiguration' kept surfacing in my research. My dictionary defines it as 'a transformation or glorification in appearance'. Glorification is the bringing of beauty, honour and happiness. Transformation literally involves a change in form, appearance and character. Our faces dominate our appearance and we spend a fortune on make-up, razors, hairdressers, dentists and surgeons to improve the way we look. Yet a glance at the average crowd in the West shows that we mostly seem pretty miserable. I believe that the majority of us need some transfiguration from within.

Transfiguration is also the vision or dream of the ideal. It is lateral thinking – thinking outside the box – that recognises the importance of people, their motivation and their spirituality. It demands a permanent spiritual step-change and a renewal of mind that I believe will allow more of us to become people of the soil, whose inner life or inner soil is grounded in the earth, who know it as their ally, and whose actions reveal their connections with the earth, with others and with the environment. Such people have inner awareness and are often called the salt of the earth.

All this is difficult to achieve when we are stuck in a system where the powers-that-be demand economic growth and are too ready to resort to the use of domination and war. This mindset often appears to be ingrained in our folklore, though closer inspection may give other insights. A traditional story from Clatt is that long ago two giants lived on Tap O' Noth and Bennachie, prominent hills located on either side of the village. Long Jock o' Noth and Long Jock o' Bennachie were both wooing the same beautiful maiden. There was no resolution so Jock o' Bennachie lobbed a huge rock high over Clatt to Jock o'Noth who was halfway down his hill. He put out his foot and stopped the stone. He chose not to fling it back and so it remains there to this day, providing shelter for hill walkers and a testament to constructive non-violence (Figure 18). No one knows who won the hand of the fair maid.

Another tale is of the rather nasty water kelpies. Legend has it that the minister of Auchindoir parish, located adjacent to Clatt, was accosted by a kelpie when crossing the burn to return to his manse. The kelpie rose up, horse-like, from under the bridge and snapped at his leg. The minister used what came most readily to hand to save himself. He stopped the water kelpie by stuffing his Bible down its throat. This could represent the forceful Christianising of the pagan world, an example, perhaps, of violence overcoming protest. However, the minister lost his Bible, maybe representing how the Bible might integrate with previous systems of spirituality to point the way to the development of more imaginative approaches to religion.

This led me to another clue to the realisation of transfiguration and spiritual change, which comes from Christian belief. Reassessing our priorities and starting afresh links us right back to a basic Christian message: 'Seek ye first the kingdom of God and his justice, and all these things (material things) will be added unto you.' To me, the kingdom of God is the spring of love within us, driving us to care and save, to create and to conserve the web of life. This is a state of spiritual awareness where all our actions are governed or ruled by love for each other and for our environment and, in non-Christian terms, is known as Love's rule. This rule or kingdom comes when we use our power, usually along with that of others, to seek for beauty and quality, from which many benefits then follow. Fritz Schumacher[5] realised that this also works in reverse. When we no longer seek first the kingdom, then what we think is ours by right is threatened and may no longer be available to us. The gradual failure of the Green Revolution in expanding food production in many parts of the world is a good example and there are others in this book. The problem is neither new nor confined to agriculture. Robert Tressell[6] realised that the great greed of Edwardian employers in the UK resulted not only in low wages and consequent social problems but also in shoddy workmanship as corners were cut to increase productivity. He described it as 'trampling the flowers to get at the worms'.

Desmond Tutu's[7] vision of the kingdom of God is people-centred: 'God calls us to be his partners to work for a new kind of society where people count; where people matter more than things, more than possessions; where human life is not just respected but positively revered; where people will be secure from the fear of hunger, from ignorance, from disease; where there will be more gentleness, more caring, more sharing, more compassion, more laughter; where there is peace and not war.' Brian McLaren[8] believes that God's kingdom advances by transformation through grace and acceptance, quietly under the surface, like seed in soil or yeast in dough. Walter Wink[9] describes the process as the creation of a domination-free order which rises quietly and imperceptibly out of the land by growth from below among the common people.

Developing a life in such an order is unlikely to be easy, though again the soil can provide a clue. Humus is described as the 'stuff' of the soil and, stemming from the Latin, the word 'human' is derived from it, as are 'humble' and 'humility'. Perhaps the way to a more sustainable lifestyle is to live humbly on the earth by cultivating our inner life or 'inner soil', so that we become resistant to degrading processes such as being stifled by domination (compaction), being addicted to consumer goods (contamination) and losing our sense of what is important in life (erosion). Then maybe 'the meek shall inherit the earth'.

Music allows us to express feelings of love and community in a way which is often more complete and more spiritually powerful than words alone. This power is well harnessed in the spiritual songs of John Bell and Graham Maule[10]. Their songbook *Love from Below*, written for the Iona Community, is 'for all the folk on the ground who have shown us love from below'. This book contains a section of songs on 'the call to care' which include titles such as 'We will not take what is not ours' and 'The greatness of the small'. They remind us of how the kingdom of God, or Love's rule, comes from below, from within thoughtful, loving people – such as carers and cleaners – who are often regarded as lowly but who transfigure our world.

Power from below should not be underestimated. I recently worked for a spell in Christchurch, New Zealand. Just before I arrived, there had been an earthquake of magnitude 7.1 which caused extensive damage to buildings, roads and the sewage system. There were many aftershocks which continued to rock and crack the buildings. After each tremor, instead of looking to the ground, I would look up to see if anything was going to drop from above. Many churches were damaged because they were old and constructed of brick. I attended one of these, Oxford Terrace Baptist Church, where the congregation was worshipping in a primary school and was thinking hard about where to go next. Unsettling can lead to growth and new life.

Fault lines

unzipped
rocks shorn
turf torn
a rumbling wave
of energy release

unzipped
mortar crumbles
chimneys tumble
pipes apart
flood the dawn

unzipped
bewilderment
suddenly aware
anxiety revealed
priorities perceived

unzipped
windows of opportunity
showing the community
harnessing
life's energies

from below

We need a transfiguration in ecology – in other words, a move to deep ecology. Alastair McIntosh[11] estimated that the Earth could sustain us for a further two billion years, but with our current rapid rate of development and societal change Ronald Wright[12] considers that we are unlikely to last beyond the end of the current century due to the impending global collapse of civilisation. We can take transfiguration literally as meaning changing the look of the whole Earth. Like many, I was enthralled back in the 1960s by the pictures of the Earth as seen from the Moon. The mottled blue and white appearance of our little globe reminded me of the Man in the Moon. I think of it as 'the Woman in the World'. The look of her face depends on our actions, mostly in soil and water management. We are called on to actively manage – to work – the soil and the Earth to transfigure it into a being of beauty and quality.

The radical change brought about by transfiguration is also needed so that we can turn our backs on domination, consumerism and destruction. One approach may be to readopt the tribal values of our ancestors, and aim for mutual support, security and safety. Fritz Schumacher's[6] vision was to develop a wisdom in which science and technology were oriented towards the holistic, the gentle, the non-violent, the elegant and the beautiful, which would lead to indivisible peace. This aligns with the aims of deep ecology[13] where the world is seen as a network or web of phenomena that are fundamentally interconnected and integrated, yet independent, and the most important values are those of conservation, co-operation, partnership and quality.

Alastair McIntosh[14] recommends a wholesale social shift towards low-input simple living in order to combat climate change and I'm sure that this would alleviate other problems too. As Gandhi put it: 'We must learn to live simply so that others may simply live.' The lifestyle of the 1900s to mid-1950s, similar to ours at Roadside, was simple but would be hard to return to after having lived a more comfortable lifestyle for so many years. In those days, we could almost exist on what others ignored, left behind or threw away. In the UK, we now throw away about one third of the food we buy, consume double the protein we need and eat too much processed food. All this would have been beyond belief in our early days at Roadside. If we could reduce our current food wastage or overconsumption, we would have less need to increase crop production.

My vision of a transfigured world is one where we care for our environment by consuming less and where we become more self-sufficient and live more simply, but with inner richness, like soils dark with organic matter. We would consume within sustainable limits and realise that enough is beautiful. As suggested by Tim Jackson[15], we could shift our emphasis away from values based on popularity, image and financial success to values of acceptance, belonging and loyalty to others. He stresses the need to appreciate much more those who do the 'real jobs' like caring for children and the elderly, those who do voluntary work and those who promote co-operation rather than competition.

We would develop intentional community just as groups of porous soils align with each other and co-exist through good interconnection of the pores. Like high-quality soil, we grow deeper and closer together, promoting 'us' rather than 'me'. In such ways Tim Jackson[15] suggests that we can prosper without growth, move to a low-carbon economy and help rid the world of poverty. 'Economic growth', the acquisition and use of things and people, is replaced with a growth of love and respect from within, an improvement in quality of life and spiritual development. This type of growth deepens our 'topsoil' by using more of what is stored deep within us, enabling our roots to spread down and outwards to intertwine with those of others in mutual understanding and love. This has a parallel in how the roots of different types of plants combine to give an increased release of nutrients from the soil around them (mycorrhizal association), whereby they feed and encourage each other. However, like good soil management, achieving a different kind of society will be a challenge.

When applied to soil management, seeking first the kingdom of God or Love's rule perhaps involves developing and sustaining soil quality (or soil health) as our first priority by supplying organic matter, and taking care of soil structure and the mineral balance so that organisms can thrive and make the topsoil rich and deep. When we do this, the soil will produce what we need sustainably. The earth is also transfigured, renewing the landscape from below: 'When you send forth your spirit they are created; and you renew the face of the earth' (Psalm 104: 30). Our imperative is not to aim at productivity for productivity's sake but to concentrate on the care and conservation of resources in ecosystems so that there is enough for all creatures as well as human beings.

Earlier I said that I was disappointed in the quality of the Insch series soil that I dug out of the field that I recently ploughed at Clatt. I then left the field and crossed the fence to a nearby natural grassy area that had not been cultivated for years. I put in my spade and there it was (Figure 17), just like at Roadside all those years ago, the beauty of the broken-up soil revealed in the rounded, nutty crumbs, warm brown in the sunlight of spring. The soil had been utterly transfigured during its period of rest by the accumulation of organic matter and the actions of worms, roots and microbes into a thing of beauty. Beauty that can feed us again and again, if we care for it.

One of the most practical approaches to linking spirituality in people with the environment was provided by St Francis of Assisi, the patron saint of ecology. His lifestyle followed that of Jesus more closely than that of many others who claimed to do so. St Francis believed that the church needed to be built spiritually on the basis of simplicity, poverty and the gospel, reflecting Jesus' extraordinary concern for the outcasts and the marginalised. St Francis lived with the poor, with few ties to material things or to power. His preaching and teaching were born from the landscape below, in the midst of the people, enabling the connection between faith and life, contemplation and action, work and celebration, thus producing a transfiguration. Leonardo Boff[16] describes the poverty identified by St Francis as not about a lack or deficiency but as something positive – a way of being in which men and women let things be as they are without dominating them. Perhaps this provides a clue as to how the ailing churches around us might 'start afresh', as suggested by one of the Clatt farmers.

Our efforts at environmental preservation and bringing equality and justice, no matter how small, are born of love and are perhaps the most important spiritual work that we do, helping to heal the world. In Christian terms, the process of healing by love is described in the Bible like the spread of the kingdom of God, similar to a seed germinating in the soil. This will eventually bring a great harvest. Taking this further, Brian McLaren believes that the core message of Christianity is less about pursuing the afterlife but more about focusing on personal, social and global transformation in this life. He compares Jesus' activity to 'excavating through layer upon layer of carpeting, plywood, ceramic tile, blacktop, gravel, trash, broken glass, and cement so that our bare feet can once again feel the cool, moist, soft soil from which we are created.' Love is then like the soil itself into which our roots can extend, allowing us to grow and be transfigured.

Part of this growth involves loving those who do not see things our way, those who oppose, thwart or refuse to offer support. After

I grew up, I came to understand the Old Man's problems and to extend sympathy. We would go out for drives in my car, which he enjoyed, and would discuss his difficulties in relating to others. He started to come round to their point of view and to realise how they felt. Bit by bit, we eventually came to an understanding so that he was loved out of his difficulties. Loving our 'enemies' is about helping them to regain their humanity. He slowly began to accept himself and others as they were. This was nothing dramatic, just an example of the power of love overcoming the love of power. Many of my heroes spoke of this love – Jesus, Martin Luther King, Gandhi. Curiously we killed them for it, but their love brought them immortality. Perhaps these deaths occurred because developing and applying such love demands a wisdom that is difficult to find and which St Paul describes in the Bible: '… it is not the wisdom that belongs to the world or the powers which rule the world – powers which are losing their power. The wisdom I proclaim is God's secret wisdom, which is hidden from mankind' (1 Cor. 2:6–7). This wisdom is found by exploring our inner lives and increasing our spiritual awareness.

Spiritual awareness can be developed by focusing on the present – developing presence. We spend so much time thinking of the past, judging things and anticipating the future. We need to find the time to be alone and silent, to be conscious of all that is around us, to feel the spirit and to become more deeply aware and present to the aliveness of life, which the jingle on Radio Caroline back in the '60s called 'loving awareness'. Desmond Tutu[7] compares engaging with the spiritual life to sitting in front of a fire on a cold day. We don't do anything, we just allow the qualities of the fire to be transferred to us and we become warm. A similar process occurs between us, God (or the encompassing Wholeness) and the environment. When we take time to be in the presence of our environment, the qualities of Wholeness are transferred to us as part of our transfiguration. If we do this outdoors, we soon become aware of things happening around us – a bird appears, we hear people or animals – and we start to reconnect with the earth. Try it. Take out your headphones and look, listen, smell and feel the environment, the Wholeness around you.

I believe that focusing on the present also brings the gift of time. Things go more slowly – not exactly at soil speed, but slow enough to make a difference. If we stop and become aware of our surroundings, time slows down. Thom Hartmann[17] suggests that when we are completely in the present, we become grounded and we wake up to and touch the true power of life. In this way we harvest the fruit of the Spirit and start to change ourselves and the world around us. This might guide us to start living towards others, as suggested by Bishop Desmond Tutu[7]. He suggests that transfiguration involves 'ubuntu', the African word for becoming aware of our own value only when we realise the value of others, looking for the light of God in them. To achieve this we do deeds of love, no matter how small – acts of kindness or mercy, or fleeting prayers and kind thoughts. If enough of us do this then our interconnection will enable a spiritual transformation that will spread across the globe. Rupert Sheldrake[18] posits that an instantaneous sharing of knowledge across widely separated locations is possible by means of 'morphic resonance'.

We are all derived from the same parent material. It is possible for us to live in harmony with those around us, in all our variability, sharing a spirit of hope, in the same way that well-interconnected porosity conducts life-giving material in the soil. Thus we can heal the void which many of us have in our lives and which is like a large, isolated soil macropore that we continually try to fill by the pursuit of money, sex or power. I experience such healing when I live with others for a few days during courses, where we share daily chores of cleaning or preparing meals along with thoughts and experiences. It is the great payback for all our struggling for wisdom and justice – a feeling of peace, personal happiness and inner security. It reduces our desire for material things and so increases sustainability. Peace of mind and spirit – true peace which invigorates life – is something that is acquired by practice and by struggle. Such opportunities allow us, according to Bernard Haisch[19], to start to transform the world from one of suspicion, intolerance and hatred to one of trust, tolerance and love.

When asked about the future of Clatt, its people expressed concerns about the construction of more houses, turning it into a 'dormitory' village where people live but make no real contribution to developing life as a community. It might be interesting to guess at an alternative future where we all aimed for transfiguration. We may return to a more natural existence, consuming much less and with greater self-sufficiency. Food, power and water would be produced locally wherever possible. These resources would be distributed equitably without profit.

In the early 1900s, the Gadie burn powered several mills, including a meal mill and sawmill. Many of the farms had dams where water stored the energy needed to thresh corn on the farm. Wind power was used to provide lighting. Now, modern technology would be exploited to allow the most efficient extraction of energy and material into useful end products. For example, a cheap nano-engine can convert bioethanol to LED lighting very efficiently. In my vision, Clatt would use energy from a mix of water, wind, solar power and biomass, although some energy would still be required from elsewhere. The demand for energy would be reduced by insulation and sharing. Although there would be less need for transport, public transport would return. Many of the houses in the village would have plots of land similar to the former crofts, where a cow is kept for milk, cheese and butter, hens are reared and food is grown. Land for vegetable production would be shared. This is already happening throughout the UK in land-share schemes. The farms would become smaller and more labour-intensive with an emphasis on biological cycling in complex rotations. Farms would begin to resemble gardens. More food crops would be grown – like oats, potatoes, vegetables and fruit. Practioners of traditional skills, such as shoemakers, tailors, doctors, teachers and grocers, might return. Advanced technology and research would be geared to achieve all these things.

Achieving such a future would be helped by human transfiguration. People who are more in touch with the earth and with their own inner lives would be more open to living with each other, rather than being isolated. As a result, many of the houses would fill with extended families and they would start to live together as a true mutually supportive community in the place that is Clatt. Although people would live and work co-operatively, their independence would still be recognised. There would still be room for people 'to get away from it all', though there would be less to get away from. We would connect back to the old culture by growing roots down to our deep Self or ecological Self (Figure 11). This growth will stimulate thinking at deeper levels and an identification with other beings, and will allow rediscovery of a community where generosity is the route to social standing, and co-operation and the movement of resources to the poor would be core values. But, best of all, living in these communities would be spiritually rewarding, bringing joy and happiness. There is no need to wait until you die to let the land or Ground of Being reclaim you. Let it happen now. The feeling of existing within the wholeness of life will bring a sense of completion, balance and healing.

All this is unlikely to occur tomorrow or the day after. But we need to start making changes here and now from where we are. Extending our roots below the grass roots of popular culture down to the parent material of the deep Self will stimulate thinking at deeper levels and allow us to express compassion and love. We can employ the same deep ecological principles[13] as in our ideal system at Clatt to start to transform our environment and our everyday existence so that we all have a part in transfiguring the face of the Earth. Harvesting the fruit of the Spirit or applying the principles of deep ecology allows us to start to identify the priorities, to realise that everyone and everything interlinks and that all creatures – not just we ourselves – have a right to live in a sustainable ecosystem.

Caring for the environment involves serious commitment. Above all, we have to understand that the only way to tackle our basic crisis of overconsumption is to consume less more equitably, to recycle as much as possible and to let go of what we no longer need. We need to encourage localisation of food and energy production to increase security of supply. There is no choice – unless we discover a miraculous new source of energy that is as versatile as oil at creating stuff. A major priority is to use our remaining oil reserves wisely, principally for producing chemicals and plastics. We can't recycle oil.

Applying deep ecology is much more than choosing lifestyle options like the purchase of an electric car or going vegetarian. Good progress towards community-oriented transfiguration is being achieved in the Transition Towns[20] movement, which is based on the principles of permaculture. However deep ecology, though intuitive and holistic, demands more. Full application of the values of conservation, co-operation, partnership and equality requires that we become less selfish and decrease our extravagant consumption. Ted Trainer[21] considers that the transition to a just and sustainable world requires a vastly reduced demand for energy. He advocates 'The Simpler Way', where communities create highly localised, zero-growth economies based on much lower resource and energy consumption and where the profit motive has largely been removed. Living economically will require a huge social change in Western society but it allows us to show solidarity with the poor and promotes social cohesion and justice. Richard Wilkinson and Kate Pickett[22] have demonstrated that happiness and trust and many other factors, from life expectancy to incidences of violence, improve for rich as well as poor as our possessions decrease and our incomes and status become more equal.

Improving our ecological relationships will also involve replacing the urge to dominate with partnership and flexibility, and returning to our basic instincts of compassion, humility and love. All are the characteristics of co-operation by which our ancestors survived for thousands of years. The lifestyle is quite similar to what we almost had 50 years ago at Roadside. We found that a low-consumption lifestyle was satisfactory, but marred by lack of co-operation, sharing and commitment – and creature comforts. Nevertheless, the peaceful spiritual revolution needed to bring in the post-consumerist social movement of 'The Simpler Way' can be joyous and make us happier and healthier.

A possible pathway to sustainability for the world as a whole, proposed by Rob Dietz and Dan O'Neill[23], is for countries to develop steady-state economies, which would be just like cultivating a good, fertile organic soil. Such an economy would involve limiting resource use, stabilising population, and distributing wealth equitably by reforming monetary systems, creating full employment and rethinking how businesses create value. Achieving this would be a huge task which would require a massive change in the powers wielded by politicians, financial institutions, corporations and religion.

From this moment forward, for us all, going green or becoming environmentally aware is not just an option, it is an imperative. Every little deed of environmental preservation, such as using food more efficiently, wearing clothes longer, saving water or reducing travel, is an important act of love in helping to transfigure our Wholeness. Prashant Vaze[24] provides lots of practical everyday suggestions.

At the centre of our transfigured world will be the wise, informed farmer who, through the fruits of research, investment, knowledge and, above all, a revitalised spiritual awareness, will start to move from agribusiness to husbandry and conservation, with a stronger link to us, the consumers. Farming will need to become smaller-scaled, mixed (i.e. including animals) and much less dependent on external energy sources – similar to the 'enlightened agriculture' described by Colin Tudge[25].

Although intensive farming based on eco-efficient principles (as described in Chapter 8) will still be required in some circumstances, this will be more dependent on good-quality soil and human labour and much less dependent on oil by-products and capital injections. Political and financial trading systems would favour smaller farmers, especially women in developing countries, leading to better rural development, greater energy and food security and greater equity. Joel Salatin[26] reckons that farmers adopting these principles may appear to be lunatics to many, but producing food based on ancient ecological principles, beautifying landscapes and keeping customers out of hospital is sheer ecstasy!

Although I'm not certain what all of the future will look like, I am sure that when we make the transfigured world, we will be standing on soil that is resilient, dark and porous, rich and deep with organic matter, wisdom and love. There will be smooth transitions between topsoil, subsoil and parent material helped by deep, wide-spreading roots. Beauty that is soil deep. We can start towards this now by walking with respect on the earth and to all its creatures. When we respect the earth we tread lightly on it. This is my great hope for the transfiguration of the Earth, and of you and me and of everyone. Look to the small and the simple, look to the beautiful, look below. From there springs peace.

Peace from below

The break in hostility
the withdrawal of troops,
peace that's fragile.
The quiet after a shouting match
the calmness of the waiting room,
peace that's uneasy.
The tranquillity of early morning
the room of a sleeping child,
peace that's temporary.
The outstretched hand to someone on the street
Listening to a wounded soul
celebrating a friend's success
sitting at a hospital bed
feeding the elderly
serving others
treading lightly
God's kingdom come
His will being done
joy at rest, love's arm enfolded.
peace
that lasts
beyond understanding.

NOTES

Chapter 1
Title verse. This and other poems attributed to Henry Ball are from Ball 1982. Other poems are by the author unless otherwise stated.

Chapter 2
[1] Lewandowski 1998.
[2] Max Purnell is a dairy farmer in New Zealand who expressed his ideas in a talk 'Knowing and growing your soil' in 2009.
[3] Dietz and O'Neill 2013.
[4] Details of the spade test are given in Ball et al. 2007.
[5] Details of the description of soil profiles by 'le profil cultural' are given in Manichon 1987.
[6] Sewall, L. The skill of ecological perception. Roszak et al 1995.
[7] Information on Hildegard von Bingen is from Hans-Peter Blume, University of Kiel.
[8] Boff 1997.

Chapter 3
[1] The title quote is from Wink 1999.
[2] Shapiro E. Restoring habitats, communities and souls. In Roszak et al. 1995.
[3] Brown 2012.

Chapter 4
[1] Boff 1997.

Chapter 5.
[1] Albert Ellis is a psychologist who pioneered the use of Rational Emotive Behaviour Therapy, see Ellis and Harper 1975.
[2] Conn, S.A. 1995. When the Earth Hurts, Who Responds? In Roszak et al. 1995.
[3] Patzel 'The soil scientist's hidden beloved: archetypal images and emotions in the scientist's relationship with soil'. In Landa and Feller 2009.
[4] McIntosh 2008a.
[5] Walsch 1999.

6   Palmer 1994.

7   The belief that violence can be overcome by violence is called the myth of redemptive violence by Wink 1999.

8   Jackson 2009.

9   Obama 2008.

10   The poem by Harry Martinson was translated from the Swedish by Robert Bly and is from Bly, R. 1975. Friends, you drank some darkness. Beacon Press, US, 267 pp.

11   Ferguson and Chater 2006.

Chapter 6

1   The opening quote is from McIntosh 2008a.

2   Schumacher 1973.

3   Information on Hildegard von Bingen is from Hans-Peter Blume, University of Kiel

4   Patzel 'The soil scientist's hidden beloved: archetypal images and emotions in the scientist's relationship with soil'. In Landa and Feller 2009.

5   Taylor 2009.

6   Lines-Kelly 2004.

7   Lewandowski 1998.

8   Sheena Blackhall's poem is available in Conn 2006.

9   Swan 1993.

10   Hartmann 1999.

11   Boff 1997.

12   Haisch 2006.

13   McIntosh 2008a.

14   McLaren 2010.

15   Abram, D. 1995. The ecology of magic. In Roszak et al. 1995.

16   Jackson 2009.

17   Hillel 1992.

18   McIntosh 2004.

19   Boff 2009.

20   Hartmann 1997.

21   Ferguson 1988.

22   McLaren 2007.

23   McIntosh 2008b.

Chapter 7

[1] The opening quote is from Howard 1947.

[2] Berry 2009.

[3] The link between the development of agriculture and inequality was observed by Wilkinson and Pickett 2009.

[4] Howard 1947

[5] Lady Eve Balfour is cited from Patzel 'The soil scientist's hidden beloved: archetypal images and emotions in the scientist's relationship with soil'. In Landa and Feller 2009.

[6] Trewavas 2004.

[7] Weckmann 2009.

Chapter 8

[1] The opening quote is from Lal 2009a.

[2] Gray, L. Shamanic counseling and ecopsychology. In: Roszak et al 1995.

[3] Allan, J. Agriculture in Aberdeenshire in the Eighteen-Sixties. In: Whiteley 1983.

[4] Shapiro, E. Restoring habitats, communities and souls. In: Roszak et al 1995.

[5] Hartmann 1999.

[6] Carter and Dale 1974.

[7] Berry 2009.

[8] Cox 2008.

[9] King 1911.

[10] The principles of permaculture are described by Whitefield 2009.

[11] Polyface farms are described by Salatin 2010.

[12] Foley et al 2005.

[13] Lal 2006.

[14] Shepherd 2009.

[15] Practical soil husbandry is described by Batey 1988.

[16] Max Purnell is a dairy farmer in New Zealand who expressed his ideas in a talk 'Knowing and growing your soil' in 2009.

[17] Lal 2009b.

[18] McIntosh 2008a.

[19] Schumacher 1973.

[20] LeChapelle and Puzey interview on deep ecology in Jensen 1995.

[21] The poem 'From oil to soil' was first published in: Ball, B.C. 2013. Spiritual aspects of sustainable soil management. In: Lal, R. and Stewart, B.A. (eds) *Principles of Sustainable Soil Management in Agroecosystems.* Advances in Soil Science book series, CRC Press, Taylor and Francis Group, Boca Raton, FL, USA, 257-284.

Chapter 9

[1] Carter and Dale 1974.
[2] Hall and Klitgaard 2011.
[3] Hartmann 1999.
[4] George MacLeod is cited in Ferguson 1988.
[5] Schumacher 1973.
[6] Tressell 2005.
[7] Tutu 2004.
[8] McLaren 2006.
[9] Wink 1999.
[10] Bell and Maule 1989.
[11] McIntosh 2004.
[12] Wright 2004.
[13] For an introduction to the principles of deep ecology, see Naess 1973 and Capra 1997.
[14] McIntosh 2008b.
[15] Jackson 2009.
[16] Boff 1997.
[17] Hartmann 1997.
[18] Sheldrake 1989.
[19] Haisch 2006.
[20] The principles of the Transition Towns movement were described by Hopkins 2008.
[21] Trainer 2010.
[22] Wilkinson and Pickett 2009.
[23] Dietz and O'Neill 2013.
[24] Vaze 2009.
[25] Tudge 2004.
[26] Salatin 2010.

FURTHER READING

Ball, B.C., Batey, T. and Munkholm, L. 2007. Field assessment of soil structural quality – a development of the Peerlkamp test. *Soil Use and Management* 23: 329-337.

Ball, H. 1982. *Reflections from Roadside*. Published by and available from the author. 92 pp.

Batey, T. 1988. *Soil Husbandry: A practical guide to the use and management of soils*. Soil & Land Use Consultants Ltd., Aberdeen, UK. 157 pp.

Bell, J.L. and Maule, G. 1989. *Love From Below*. Volume 3 of Wild Goose Songs. Iona Community, Glasgow. 144 pp.

Berry, W. 1996. *The Unsettling of America: Culture and agriculture*. 3rd Edition. Sierra Club Books. 234pp (cited by Lines-Kelly, 2004).

Berry, W. 2009. *Bringing It to the Table: On farming and food*. Counterpoint, Berkeley, USA. 234 pp.

Boff, L. 1997. *Cry of the Earth, Cry of the Poor*. Orbis, New York. 242 pp.

Boff, L. 2009. *Eco-simplicity*. Http://www.leonardoboff.com/site-eng/lboff.htm

Brown, D. 2004. *The Da Vinci Code*. Bantam Books, UK. 583 pp.

Brown, L.R. 2012. *Full Planet, Empty Plates*. Norton, New York, USA. 141 pp.

Capra, F. 1997. *The Web of Life: A new synthesis of mind and matter*. Harper-Collins, London, UK. 320 pp.

Carter, V.G. and Dale, T. 1974. *Topsoil and Civilization* (Revised edition). University of Oklahoma Press, USA. 292 pp.

Conn, S. (ed.) 2006. *100 Favourite Scottish Poems*. Luath Press, Edinburgh.

Cox, S. 2008. *Ending 10 000 Years of Conflict between Agriculture and Nature*. Institute of Science in Society Report, London UK. 7 pp.

http://www.i-sis.org.uk/Ending10000YearsOfConflict.php

Dietz, R. and O'Neill, D. 2013. *Enough is Enough.* Earthscan/Routledge, Abingdon, UK. 240 pp.

Ellis, A. and Harper, R.A. 1975. *A Guide to Rational Living.* Third edition. Wilshire Book Company, USA. 283 pp.

Ferguson, R. 1988. *Chasing the Wild Goose.* Fount Paperbacks, London. 208 pp. New edition 1998, Wild Goose Publications, Glasgow.

Ferguson, R. and Chater, M. 2006. *Mole Under the Fence. Conversations with Roland Walls.* Saint Andrew Press, Edinburgh. 191 pp.

Foley, J.A., Defries, R., Asner, G.P. et al. 2005. Global Consequences of Land Use. *Science* 309: 570-573.

Haisch, B. 2006. *The God Theory: Universes, zero-point fields and what's behind it all.* Weiser Books, San Francisco, USA. 157 pp.

Hall, C.A.S. and Klitgaard, K.A. 2011. *Energy and the Wealth of Nations: Understanding the biophysical economy.* Springer, New York, USA. 407 pp.

Hartmann, T. 1997. *The Prophet's Way: Touching the power of life.* Hodder Mobius, London, UK. 335 pp.

Hartmann, T. 1999. *The Last Hours of Ancient Sunlight. Waking up to personal and global transformation.* Hodder and Stoughton, London. 314 pp.

Hillel, D. 1992. *Out of the Earth. Civilization and the life of the soil.* University of California Press, Los Angeles, USA. 321 pp.

Hopkins, R. 2008. *The Transition Handbook: From oil dependency to local resilience.* Green Books, Totnes, Devon. 240pp.

Howard, A. 1947. *The Soil and Health: A study of organic agriculture.* The Devin-Adair Company, New York, USA. 307pp.

Jackson, T. 2009. *Prosperity Without Growth? The transition to a sustainable economy.* Sustainable Development Commission Report, UK. 133 pp.

Jensen, D. 1995. *Listening to the Land. Conversations about Nature, culture and Eros.* Sierra Club Books, San Francisco, USA. 328 pp.

King, F.H. 1911. *Farmers of Forty Centuries: Permanent agriculture in China, Korea and Japan.* Rodale Press, Pennsylvania. 441 pp.

Lal, R. 2006. Managing Soils to Feed a Global Population of Ten Billion. *Journal of Science, Food and Agriculture* 86: 2273-2284.

Lal, R. 2009a. Soils and World Food Security. *Soil & Tillage Research* 102: 1-4

Lal, R. 2009b. Technology Without Wisdom. In: Lichtfouse, E. (ed.) *Organic Farming, Pest Control and Remediation of Soil Pollutants.* Springer, New York, USA, pp. 11-14

Lal, R. 2009c. Tragedy of the Global Commons: Soil, Water and Air. In: Lichtfouse, E. (Ed.). *Climate Change, Intercropping, Pest Control and Beneficial Microorganisms.* Sustainable Agriculture Reviews 2, Springer, New York, USA, pp. 9-11.

Landa, E.R. and Feller, C. (eds.) 2009. *Soil and Culture.* Springer. 524 pp.

Lewandowski, S. 1998, Wild Soils. *The Crooked Lake Review,* Summer issue. Http://www.crookedlakereview.com/articles/101_135/108summer1998/108 lewandowski.html

Lines-Kelly, R. 2004. *Soil: Our Common Ground – A humanities perspective.* National Soils Conference, Sydney.

Lumsden, J.A. 1995. *A Short History of the Parish of Clatt in the Twentieth Century.* Published by the author. 76 pp.

Manichon, H. 1987. Observation Morphologique de l'État Structural et Mise en Evidence d'Effects du Compactage des Horizons Travaillés. In: Monnier, G., Goss, M.J. (eds.). *Soil Compaction and Regeneration.* Balkema, Rotterdam, The Netherlands, pp. 35-52.

McIntosh, A. 2004. *Soil and Soul. People versus corporate power.* Aurum Press, London. 326 pp.

McIntosh, A. 2008a. *Rekindling Community: Connecting people, environment and spirituality.* Schumacher Briefing 15, Green Books for the Schumacher Society, Bristol. 112 pp.

McIntosh, A. 2008b. *Hell and High Water: Climate change, hope and the human condition.* Birlinn, Edinburgh. 289 pp.

McLaren, B.D. 2006. *The Secret Message of Jesus. Uncovering the truth that could change everything.* Thomas Nelson, Nashville, USA. 263 pp.

McLaren, B.D. 2007. *Everything Must Change: Jesus, global crises and a revolution of hope.* Thomas Nelson, Nashville, USA. 327 pp.

McLaren, B.D. 2010. *Naked Spirituality.* Hodder and Stoughton, London. 352pp.

Naess, A. 1973. The Shallow and the Deep, Long-range Ecology Movement. A summary. *Inquiry* 16: 95-100.

Obama, B. 2008. *Dreams From My Father.* Canongate, Edinburgh, UK. 442 pp.

Palmer, H. 1994. *Living Deliberately: The discovery and development of Avatar.* Star's Edge International, USA. 127 pp.

Roszak, T., Gomes, M.E. and Kanner, A.D. 1995. *Ecopsychology: Restoring the earth, healing the mind.* Sierra Club Books, University of California Press. 338 pp.

Salatin, J. 2010. *The Sheer Ecstasy of Being a Lunatic Farmer.* Polyface, Virginia, USA. 315 pp.

Schumacher, E.F. 1973. *Small is Beautiful. A study of economics as if people mattered.* Vintage, London. 260 pp.

Sheldrake, R. 1989. *The Presence of the Past: Morphic resonance and the habits of nature.* Vintage Books, New York, USA. 416 pp.

Shepherd, T.G. 2009. *Visual Soil Assessment. Volume 1. Field guide for pastoral grazing and cropping on flat to rolling country.* 2nd Edition. Horizons Regional Council, Palmerston North, New Zealand. 106 pp.

Swan, J.A. 1993. *The Power of Place. Sacred ground in natural and human environments.* Gateway Books, UK. 365 pp.

Taylor, B.B. 2009. *An Altar in the World: Finding the sacred beneath our feet.* Canterbury Press, UK. 216 pp.

Trainer, T. 2010. *The Transition to a Sustainable and Just World.* Envirobook, Sydney. 330 pp.

Tressell, R. 2005 (unabridged edition with notes). *The Ragged-Trousered Philanthropists.* Oxford World Classics, Oxford, UK. 649 pp.

Trewavas, A. 2004. A Critical Assessment of Organic Farming-and-Food Assertions with Particular Respect to the UK and the Potential Environmental Benefits of No-Till Agriculture. *Crop Protection* 23: 757-781

Tudge, C. 2004. *So Shall We Reap.* Penguin, UK. 451 pp.

Tutu, D.M. 2004. *God Has a Dream. A vision of hope for our time.* Random House, London. 134 pp.

Vaze, P. 2009. *The Economical Environmentalist. My attempt to live the low-carbon life and what it cost.* Earthscan, London, UK. 350 pp.

Walsch, N.D. 1999. *Conversations with God. Books 2 and 3. An uncommon dialogue.* Hodder and Stoughton, London. 262 pp and 392 pp.

Weckmann, J. 2009. *Organic, Local and Fair – One company's attempt to tackle environmental and climate challenges. Abstract of speech for workshop 'Global challenges – organic approaches: organic food and farming in times of climate change, biodiversity loss and global food crisis'.* 2nd European Organic Congress, Brussels, pp. 39-41

White, L. 1967. The Historical Roots of our Ecological Crisis. *Science* 155 (3767), 1203-1207.

Whitefield, P. 2009. *Permaculture in a Nutshell* (5th Edition). Permanent Publications, UK. 84 pp.

Whiteley, A.W.M. (ed.) 1983. *Bennachie Again.* Bailies of Bennachie, Aberdeen, UK, pp. 54-63.

Wilkinson, R. and Pickett, K. 2009. *The Spirit Level: Why equality is better for everyone.* Penguin, UK. 375 pp.

Wink, W. 1999. *The Powers That Be: Theology for a new millennium.* Galilee, Random House, NY, USA. 224 pp.

Wright, R. 2004. *A Short History of Progress.* Canongate, Edinburgh, UK. 211 pp.